Feeling Good
and
Doing Better

CONTEMPORARY ISSUES IN BIOMEDICINE, ETHICS, AND SOCIETY

Feeling Good and Doing Better, edited by **Thomas H. Murray, Willard Gaylin,** and **Ruth Macklin,** *1984*

Ethics and Animals, edited by **Harlan B. Miller** and **William H. Williams,** *1983*

Profits and Professions, edited by **Wade L. Robison, Michael S. Pritchard,** and **Joseph Ellin,** *1983*

Visions of Women, edited by **Linda A. Bell,** *1983*

Medical Genetics Casebook, by **Colleen Clements,** *1982*

Who Decides?, edited by **Nora K. Bell,** *1982*

The Custom-Made Child?, edited by **Helen B. Holmes, Betty B. Hoskins,** and **Michael Gross,** *1981*

Birth Control and Controlling Birth, edited by **Helen B. Holmes, Betty B. Hoskins,** and **Michael Gross,** *1980*

Medical Responsibility, edited by **Wade L. Robison** and **Michael S. Pritchard,** *1979*

Contemporary Issues in Biomedical Ethics, edited by **John W. Davis, Barry Hoffmaster,** and **Sarah Shorten,** *1979*

Feeling Good and Doing Better

Ethics and Nontherapeutic Drug Use

Edited by

*Thomas H. Murray, Willard Gaylin,
and Ruth Macklin*

Humana Press • Clifton, New Jersey

Library of Congress Cataloging in Publication Data

Main entry under title:

Feeling good and doing better.

    Includes bibliographical references and index.
    1. Drug abuse—United States—Addresses, essays,
lectures.   2. Psychotropic drugs—United States—
Addresses, essays, lectures.   I. Murray, Thomas H.
II. Gaylin, Willard.   III. Macklin, Ruth.
HV5825.F44   1984   363.4'5'0973   84–4552
ISBN 0-89603-061-X

© 1984 The Humana Press Inc.
Crescent Manor
PO Box 2148
Clifton, NJ 07015

Printed in the United States of America

Preface

The place of drugs in American society is a problem more apt to evoke diatribe than dialog. With the support of the National Science Foundation's program on Ethics and Values in Science and Technology, and the National Endowment for the Humanities' program on Science, Technology, and Human Values,* The Hastings Center was able to sponsor such dialog as part of a major research into the ethics of drug use that spanned two years. We assembled a Research Group from leaders in the scientific, medical, legal, and policy communities, leavened with experts in applied ethics, and brought them together several times a year to discuss the moral, legal and social issues posed by nontherapeutic drug use. At times we also called on other experts when we needed certain issues clarified.

We did not try to reach a consensus, yet several broad areas of agreement emerged: That our society's response to nontherapeutic drug use has been irrational and inconsistent; that our attempts at control have been clumsy and ill-informed; that many complex moral values are entwined in the debate and cannot be reduced to a simple conflict between individual liberty and state paternalism.

Of course each paper should be read as the statement of that particular author or authors. The views expressed in this book do not necessarily represent the views of The Hastings Center, the National Science Foundation, or the National Endowment for the Humanities.

We would be ungenerous if we did not express our thanks to the research assistants who worked on this project—Carola Mone and Robert Kinscherff.

*The official title of the project was "The Regulation of Drugs: Medical and Health Models." It was NSF-EVIST grant #OSS-06981.

Their efforts to keep us well-informed deserve hearty acclaim. Likewise, we owe much to our administrative assistant and chief typist, Eva Mannheimer, and to Marie Grilli O'Mara.

Thomas Murray
Willard Gaylin
Ruth Macklin

CONTENTS

Preface v

Feeling Good and Doing Better 1
by Willard Gaylin

Part I: Social and Political Aspects

Drug Abuse Policies and Social Attitudes to
Risk Taking 13
by James B. Bakalar and Lester Grinspoon

The Social Dilemma of the Development of a
Policy on Intoxicant Use 27
by Norman E. Zinberg

Controlling the Uncontrollable 49
by John P. Conrad

The State's Intervention in Individuals' Drug
Use: A Normative Account 65
by Robert Neville

Part II: Pleasure and Performance

The Use of Drugs for Pleasure: Some Philo-
sophical Issues. 83
by Dan W. Brock

Drugs, Sports, and Ethics 107
by Thomas H. Murray

Part III: Privacy, the Constitution, and Drug Use

Implications of the Constitutional Right of Privacy for the Control of Drugs:
An Introduction..............................129
by *Robert L. Schwartz*

Using and Refusing Psychotropic Drugs 157
by *Nancy K. Rhoden*

Part IV: Drugs, Models, and Moral Principles

Doctors, Drugs Used for Pleasure and Performance, and the Medical Model 175
by *Robert Michels*

Drugs, Models, and Moral Principles..........187
by *Ruth Macklin*

Index......................................215

Contributors

JAMES BAKALAR • *Massachusetts Mental Health Center, Boston, Massachusetts*

DAN W. BROCK • *Department of Philosophy, Brown University, Providence, Rhode Island*

JOHN CONRAD • *Sacramento, California*

WILLARD GAYLIN • *The Hastings Center, Hastings-on-Hudson, New York*

LESTER GRINSPOON • *Massachusetts Medical Health Center, Boston, Massachusetts*

RUTH MACKLIN • *Albert Einstein College of Medicine, Bronx, New York*

ROBERT MICHELS • *Department of Psychiatry, Cornell University Medical College, New York Hospital, New York, New York*

THOMAS MURRAY • *Institute of Society, Ethics, and the Life Sciences, The Hastings Center, Hastings-on-Hudson, New York*

ROBERT NEVILLE • *Department of Philosophy and Religion, SUNY-Stony Brook, Stony Brook, New York*

NANCY K. RHODEN • *The Ohio State University, College of Law, Columbus, Ohio*

ROBERT L. SCHWARTZ • *University of New Mexico School of Law, Albuquerque, New Mexico*

NORMAN ZINBERG • *Department of Psychiatry, The Cambridge Hospital, Cambridge, Massachusetts*

Feeling Good and Doing Better

An Introduction

Willard Gaylin

We are—it seems—finally at that long anticipated threshold of a scientific understanding of human behavior. Although this vast area of unexplored territory in human physiology remains essentially still inviolate, we are beginning our early exploratory thrusts.

The chief advances in recent years have been in the understanding and development of mood-altering drugs. Here, significant new findings in the fields of neurobiology and psychopharmacology offer great promise in understanding mental illness, mental retardation, drug addiction, and other health problems. Beyond that, however, they may offer an opportunity to explore powerful nonmedical aspects of the human condition: variable capacities for pain, for pleasure, for learning.

These new developments have been greeted with general excitement in the academic community. To the chagrin of many, however, this research, like that in genetic engineering, seems to arouse as much discomfort as it does satisfaction. The Hosannahs were there, but so was the hand-wringing. It is a troublesome development that has to be anticipated and dealt with. It is part of what I have referred to as the "Frankenstein Factor."[1] Research that changes or controls "the nature of our species" or allows for any "mechanical" influencing of human behavior will almost inevitably be received with more fear than other research that may be riskier for the individual and more dangerous to the species. A mechanical heart excites the popu-

1

lation, because any device that saves and extends human life
aggrandizes not only the discoverer, but also the species. What
an extraordinary animal is homo sapiens! Obviously, such control over death sets us apart from the helplessness of the general animal host. But mechanically influencing behavior is another matter. We are the researcher, but we are also the
subject. The manipulation of human behavior reasserts human
kinship with the pigeon, the rat, and the guinea pig.

The second element in the Frankenstein Factor is that, perversely, the more technological the control devices—the more
mechanical the method—the scarier it all seems to the general
public. I say "perversely" because obviously low-technology
methods of influencing human nature are much more
exploitable by a corrupt government or segment of society than
high technology. It is certainly easier to control values—and
politics—by manipulation of the educational systems, control
of television, revision of history, restriction of flows of information, and indoctrination through organized structures of the
state than by planting electrodes in the heads of dissidents. The
Soviet Union today is a testimony to such low-technology manipulation. Nonetheless, normal sensory inputs seem "natural" to us, whereas artificial inputs such as drugs, electrodes,
psychosurgery seem "unnatural."

The natural/unnatural polarity is one that we have clung
to, and for some good reasons, but not all are necessarily
sound, nor still operative. Certainly, in terms of the near future, the natural input is the more corruptible. Then again,
what is natural? It is our nature to change our nature. We are
constantly altering our processes of behavior and our modes of
perception. The border between the natural and unnatural has
always been blurred, and recent research has compounded the
problems. It is important that irrational fears be allayed as new
research probes our potential for understanding behavior,
facilitating performance, easing anguish, and the like. Some
valid distinctions can surely be made. It is time to reexamine
some of the past criteria for deciding right and wrong in modi-
fying our behavior.

The Medical Model

In the past, the explicit or implicit adherence to the "medical mode" was the primary means of distinguishing between

the socially acceptable and unacceptable use of drugs. The medical model, specifically the "disease" model, assumed an underlying organic cause for mental disorders or deviant behavior. The tacit guidelines of the past operated under this assumption to establish what was legitimate and illegitimate when using mind-altering drugs. Although these guidelines were rarely stated explicitly, they seem to have operated as follows:

1. Drugs are an acceptable substance for the curing of disease and relieving pain.
2. Following from this principle, drugs are acceptable if their purpose is to bring a person's physiological or behavioral function up to medically determined levels of normalcy, i.e., as long as you are reestablishing a normal level no moral issues seem to be involved.
3. Drugs are not considered acceptable, however, if they are serving purely recreational purposes, or if they seem to move beyond replacement into enhancement or improving performance or behavior.

There have always been paradoxes in our applications of this model. For some reason, we have been less distressed by drugs that bring us down than drugs that perk us up. In this "pill-popping" society of ours, tranquilizers such as Valium, sold in incredible numbers, seem quite acceptable at a time when we are frightened and horrified at the use of drugs to induce pleasurable states of consciousness, for example, marihuana. This is not to endorse the use of either of these drugs; each has their problems and, indeed, the danger is enhanced in the case of marihuana by our knowledge that it was used extensively by the young and was destroying their adolescence. Nonetheless, a contradiction does exist, and Gerald Klerman, a leading psychiatrist, referred to this dichotomy as "pharmacological Calvinism" versus "psychotropic hedonism."[2]

Current research in the field of neurobiology and psychopharmacology has forced us to question the soundness of the three traditional "guidelines." Frontier research on a class of drugs known as endorphins—research described by leading experts as "the hottest research in neurobiology"—has muddied the water further. This research has shown an intriguing interrelationship between addiction, pain, and schizophrenia, but beyond the specific medical area, it has found that endorphins tend to enhance attention and improve memory in a research population of mentally retarded.

Perhaps the most striking fact about these substances is their close relationship to the "artificial" opiates, morphine, and heroin. The work on the endorphins requires that we reexamine the distinction between endogenous and exogenous, between artificial and natural, further undermining some of the fragile distinctions that have supported us in the past.

The project was undertaken to examine some of these traditional assumptions that underlay the social control and legal regulation of drugs. The unifying theme running through the inquiry is the basis for distinguishing between drugs that "normalize" mood or behavior and drugs that "optimize" emotions or performance.

The intention of this study was to reexamine the manner in which the medical model has informed our understanding of legitimate and illegitimate uses of drugs, and to inquire whether the traditional medical model was truly the most appropriate one for determining the conditions under which drugs should be dispensed, regulated, and controlled. We structured the work under two major headings.

Inquiry I: Endorphin Research and Its Implications

The discovery of the first endorphins—the enkephalins—in 1975 by Drs. Hughes and Kosterlitz marked a new era in psychopharmacology.[3] "Endorphin" is the name given to a class of endogenous peptides that meet the two major tests for opioid drugs.[4] Investigation of the neurobiological function of the endorphins has been intense in two areas of medical research: one concerned with the neurophysiological roots of mental illness; and the other concerned with the mechanisms of pain perception and its alleviation through the opioid drugs.

Not only do the endorphins seem to explain the operation of the opiate receptors in the CNS pathways and the mechanism of the function of opiates in relieving pain, but they also might explain the addictive properties of the opioid drugs. There is some evidence indicating that excesses and deficiencies of endorphins are involved in certain mental illnesses, specifically schizophrenia and depression.[5] Clinical observations of heroin addicts suggest that some of these individuals use opiates as a form of self-medication.[6]

These findings are suggestive for a number of questions that relate to the broad, unifying themes of this project. There has been much debate over whether drug addiction should properly be considered a deficiency disease. Dole and Nyswander, who developed methadone maintenance, made explicit reference to addiction as a "deficiency disease," drawing on the analogy with diabetes and insulin deficiency.[7] This was prior to the links that current research reveals between drug addiction and endorphin levels. In the ongoing debates, one side argues that the causes of addiction are to be found in social circumstances and that even the practice of methadone maintenance is an inappropriate intrusion of the medical model into an arena that needs to be approached by changing the social order.[8] On the other side, it is pointed out that those who suffer from addiction to drugs have many of the same qualities as those who suffer from unquestionable diseases: they are unable to rid themselves of their conditions by efforts of their will; they suffer severe physical symptoms during withdrawal; many revert to addiction even after being "cured," in the way chronic disease sufferers (such as asthmatics) have repeated occurrences.[9] Does the research on endorphins lend weight to the viewpoint that treats addiction as a deficiency disease? Do the clinical observations of addicts who self-medicate imply that these are cases of "replacement therapy" rather than of the use of drugs for mood enhancement, escape, or inducement of pleasurable states. If so, is this behavior that falls fully within the traditional medical model? Does the evidence regarding connections between endorphin levels and mental illness lend additional confirmation to the hypothesis that schizophrenia and depression have demonstrable organic bases and are not therefore, simply "problems in living" as Thomas Szasz and other opponents of the medical model have argued?

Although the results of this neurobiological research by individuals such as Dr. Solomon Snyder at Johns Hopkins and the late Dr. Nathan Kline are not yet conclusive, it seems clear that the endorphins are causally involved in human behavior. The knowledge that the human brain manufactures its own opiates—even if we do not yet know to what specific end—demands a reevaluation of our current attitudes toward the endorphins' close relatives: the "artificial" opiates. If endorphin deficiencies or excesses are causally related to deviant behavior or abnormal emotional states of various sorts, how does this bear on the traditional way of looking at normalizing versus optimizing behavior or emotions? This consideration, together with the findings related to memory and learning capacities, tends to blur the distinction between substances thought of as normalizing (and thus acceptable under a traditional medical model) and those that have in the past been considered as life-enhancing or pleasure-producing (and thus unacceptable under the medical model). Although this traditional distinction may no longer hold, the question still remains about the appropriate uses of such substances. If a medical model provides no basis for answering that question, what alternatives are available?

One task was to ascertain how the word "normal" was being used in connection with endorphin excesses and deficiencies. Was the term being used in a statistical sense? In a way derivable from knowledge of the correct biochemical function of human beings? Or was it used in a normative sense, such that endorphin excesses or deficiencies are considered "abnormal" if the individual exhibits deviant behavioral or affective traits? In this sense, does the disorder become defined in terms of deficiency and surplus? What are the implications of such efforts to redefine behavioral problems?

Inquiry II: Performance Enhancers

Most of the research on performance enhancers has been conducted in connection with athletic performance. It is well known that athletes (and horses) must often submit urine or blood samples to demonstrate that their performance has not been improved by ingestion of drugs. Research on the effects of the amphetamines on swimmers has gone on for decades.[10] Athletes in competitive performances found to have taken drugs believed to enhance their abilities are disqualified. Protests have been registered against infusions of extra blood on the part of athletes. Blood infusion is another form of performance enhancer, since the extra oxygen provides an athlete with more endurance.[11] A similar example is direct inhalation of oxygen—a practice carried on by football players during the course of the game.

What is the operative assumption in the practice of banning performance enhancers in athletics? This is not a field in which the medical model has traditionally held sway. One of our tasks was to explore the rationale for the prohibition of drugs that may enhance performance in athletics. Though not subject to official medical regulation and control, a code of norms nonetheless governs this field, as in other competitive endeavors. If performance enhancers were made available to all who wanted to use them, what arguments could be offered against this attempt to optimize performance in competitive settings?

A central aspect of this inquiry was a study of the standard ways of distinguishing normalizing from optimizing behavior, mood, and performance. It has traditionally been considered acceptable to use drugs for the purpose of bringing people who are naturally deficient in some way to a normal state. One illustration is the use of insulin for diabetics—a substance normally manufactured by the human body, but lacking in those suffering from diabetes. If it is a legitimate use of drugs to use them as "replacement therapy," why should there be social and ethical disapproval of using drugs (whether naturally occurring or synthesized) to optimize or to enhance abilities or performance? What rationale lies behind the approval of normalization of human functions, but the prohibition of their optimization? Why should performance enhancers be socially and ethically permitted when the aim is to get *sub*normal individuals to perform better, but not when the purpose is to aid *normal* individuals to perform even better?

A familiar instance of the use of drugs to normalize behavior or performance is that of administering amphetamines or Ritalin to hyperactive youngsters, who typically have a short attention span. Questions have been raised about whether the medical model provides the most appropriate way of conceptualizing and treating behavioral aberrations of this sort. Controversy continues to rage about the wisdom of treating hyperactivity in children with drugs at all, as well as the propriety of administering treatment or doing research in elementary school settings.[12] This use of drugs has been defended on the grounds that children can, as a result, learn better.

The same questions arise in all areas where performance enhancers are at issue. In 1973 the *New York Times* reported that a physician had been giving injections of stimulant drugs to persons who were well known in public life.[13] It is clear that what this physician was doing was in violation of existing laws. But leaving aside for a moment the illegality of his acts, what—if anything—is morally wrong with offering or self-administering substances believed to enhance cognitive or other work? Does the bias against the use of substances with a potential for improving performance that already falls within the range of normalcy stem from the (medical model) distinction between optimizing and normalizing? Or do other factors contribute to the social and legal disapproval of physicians prescribing or using stimulant drugs for purposes that are not strictly medical?

Put succinctly, what is it that makes drugs different? Is there a legitimate distinction to be made between influencing behavior by normal sensory inputs as opposed to synthesized chemicals in the form of drugs? Consider the use of learning devices during sleep or behavior modification (which also bypasses the rational processes) with the aim of increasing learning capacity. Why should these be considered acceptable means of enhancing an individual's cognitive ability or performance, whereas drugs are morally and legally proscribed for this purpose? Even more to the point, caffeine is perhaps the most widely used stimulant and is a wholly acceptable substance in the workplace—often available without charge to employees. What makes stimulant drugs different? Does the difference stem—once again—from factors related to the medical model: food substances, such as coffee, do not lie within the province of medical prescription and control, through pharmaceutical products do?

In surveying drug regulations and unofficial norms governing the use of amphetamines and other substances that improve performance, this inquiry served as a case study for the larger theme of the project: the distinction between normalizing and optimizing. At the same time, it addressed the different ethical, social, and legal implications that both the medical and health models have for the regulations and control of drugs that can improve capacity or performance.

Of course, all of us in planning this project were aware of the extraordinary abuse of drugs and the vast range of potential harm they can inflict. We are not attempting to justify the use of drugs, but rather to understand the rationale that distinguished between drugs and other forms of help—or for that matter, abuse. Certainly, exercise, training, and discipline are useful and respected modes of performance enhancement, but an examination of the techniques used on children training to be figure skaters or ballet dancers show that the abuse to the body, and the crippling of normal physiological development and potentials were certainly as great in the areas of exercise, enforced diet, and the like, as through the use of steroids and drugs. Yet this severe training did not seem to compel the attention or command the outrage justified. Somehow or other we are more sensitive to abuse by drugs.

This is another manifestation of our bias against the "artificial" at a time when the definitions of natural and artificial are becoming more obscured.

Because of our concern about potential abuses from multiple sources, it seemed crucial that we reexamine the fundamental assumptions on which we base our control of drugs. We fully anticipated that we might conclude that regulations on drugs were not excessively restrictive, merely irrationally grounded. We might conclude that regulation of other forms of activities might be indicated, particularly where the protection of minors was a factor. Most important, where it was evident that past assumptions ought be tested in the light of current knowledge, that an attempt be made to dispel the biases of the Frankenstein Factor, that we must recognize areas of potential danger, and that controls wherever needed be based on understanding rather than irrational fear. It is to serve these purposes that this work was undertaken and is presented, in part, here.

Notes and References

[1]Willard Gaylin, "The Frankenstein Factor," *New England Journal of Medicine*, September 22, 1977, pp. 665–666.

[2]Gerald L. Klerman, "Psychotropic Hedonism vs. Pharmacological Calvinism," *Hastings Center Report*, 2:4, September 1972, pp. 1–3.

[3]—"Human Beta-Endorphin: the Real Opium of the People," *British Medical Journal*, July 15, 1978.

[4]This means that (1) they produce the same pharmacological effects as the opioid agonists (e.g., morphine), and that (2) they are antagonized by naloxone.

[5]Karl Vereby et al., "Endorphins in Psychiatry: an Overview and a Hypothesis," *Archives of General Psychiatry*, 35:7, July 1978.

[6]Karl Vereby, op. cit, p. 880

[7]Vincent P. Dole and Marie Nyswander, "A Medical Treatment for Diacetylmorphine (heroin) Addiction," *Journal of the American Medical Association* 193, August 23, 1965, pp. 646–650.

[8]Edward Jay Epstein, "Methadone: the Forlon Hope" *The Public Interest,* Summer 1974, pp. 3–23; Dorothy Nelkin, *Methadone Maintenance: A Technological Fix,* New York: George Braziller, 1973; Ronald Bayer, "Methadone Under Attack" *Contemporary Drug Problems,* Fall 1978, pp. 367–400.

[9]Vincent P. Dole and Marie Nyswander, "Heroine Addiction—a Metabolic Disorder," *Archives of Internal Medicine,* **120,** July 1962.

[10]Haldi and W. Wynn, "Action of Drugs on Efficiency of Swimmers," *Research Quarterly* **17,** pp. 96–101; G. M. Smith and H. K. Beecher, "Amphetamine Sulphate and Athletic Performance: I. Objective Effects," *Journal of the American Medical Association* **170,** 1959, pp. 542–557.

[11]Melvin Williams, "Blood Doping: Does it Really Help Athletes?" *Physician and Sports Medicine,* January 1975, pp. 52–55.

[12]"Robert Neville: What moral difference does it make if there is a biological impairment? What happens if you have a kid who behaves in a way that lots of people don't like? And you find that there's a drug that would cure him, and he and everybody else wants to take the drug that would cure him, even though there is no biological impairment? Hein: Not cure him. Control his behavior." In: MBD, Drug Research and the Schools, special supplement *Hastings Center Report* **6,** June 1976.

[13]*The New York Times,* April 19, 1973, pp. 1–25.

Part I

Social and Political Aspects

Drug Abuse Policy and Social Attitudes to Risk-Taking

James B. Bakalar and Lester Grinspoon

Some kinds of risk-taking in the pursuit of pleasure or the improvement of work performance are considered socially and legally acceptable. Drug use is not, except for alcohol, tobacco, and caffeine. This is certainly not because the dangers of drugs have been carefully compared with other kinds of danger, and not even because the risks of legal drug use have been carefully balanced against the benefits, or against the cost of enforcing punitive laws. Ironically, if we did those things, alcohol and tobacco might be the first drugs banned. But this kind of reckoning is regarded, in effect, as too difficult and uncertain. Instead, social and legal policy is guided by certain unconscious analogies, whereas others are ignored or rejected. The analogies we actually use are not entirely misleading, but they are partial; in this case they justify severe social disapproval and legal restraints that we would not tolerate for equivalent dangers not produced by drugs. It is useless to protest against this in the name of abstract consistency and rationality; the current drug control arrangements and their justifying analogies have the weight of all twentieth century history behind them, and they are similar everywhere in the world. They are the accepted and perhaps inevitable response, in our time, to a social need for classification and control of a complicated and ambiguous phenomenon. But at least we should allow ourselves to be more aware of what we are doing and describe it more candidly.

13

Consider a typical article on the dangers of what is collo-
quially known as getting high. The alarming growth of interest
in this behavior is pointed out, its social roots are examined,
the epidemiology of the habit is analyzed, and the motives of
the young people involved are discussed: some are depressed,
others bored; some want escape, adventure, or a way of testing
themselves in the search for identity; a few have personality
disorders, and more suffer from feelings of inadequacy. The influence of parents, peers, and social conditions is discussed.
The sometimes remarkably elaborate paraphernalia used and
the proliferation of shops displaying and selling these items
without interference from the law are mentioned and deplored.
The abnormal mental and physical states produced by the habit
are analyzed; serious accidents and even death are among the
dangers emphasized. The possibility of psychological dependence is considered, and the article calls for more effective law
enforcement and a re-examination of the spiritual and moral
condition of the society that has produced the epidemic.

It all sounds unremarkable; the ideas and vocabulary are
very familiar. But a few years ago a satirical piece with this
theme was published under the title "Alpinism: The Social, Scientific and Treatment Aspects of Getting High and Its Prohibition."[1] The point, of course, is that much of the descriptive language in writing about drug use applies just as well to
mountain climbing, although the usual anxious soul-searching
and earnest recommendations sound solemnly absurd if
transferred from drugs to mountain climbing. For mountain
climbing you may also substitute sailing small boats, boxing, or
high school football, which produces several dozen cases a year
of permanent quadriplegia (paralysis from the neck down).[2]
Obviously no one is calling for high school football to be outlawed. It would be silly to suggest that drug use is just like
these other pursuits and should be subject to similar rules, but
the analogy is not entirely irrelevant either. As the satire
implies, we tend to deny similarities between drug taking and
other voluntary activities that involve some risk; clearly we do
not want to find ourselves comparing the effects of marihuana
on high school students with the effects of football.

One reason is an implicit judgment about the potential
value of the activity. Mountain climbing may be dangerous, but
its pleasures are virtuous ones; it provides an opportunity for
physical exercise, adventure, achievement, and the enjoyment
of natural beauty. In making public policy, the illness caused

by enjoyable but unhealthy eating habits or the accidents suffered by careless mountain climbers and motorcycle riders are balanced against the satisfaction of human needs and desires that these activities are assumed to provide. But drug use for anything but the treatment of disease is not regarded as satisfying a legitimate need or desire. There is some underground lore about the alleged beneficial effects of illict drugs—even some claims about the naturalness of a desire to alter consciousness—and there is also a more respectable lore about the virtues of alcohol. But it all tends to be nervously jocular or embarrassed and has little effect on public policy. Permitting drug use is sometimes defended in the name of individual freedom, but rarely on the ground that there is any good in it.

By assuming that drug taking has no positive value, we implicitly make use of second analogy, which presents drug control as a form of consumer safety legislation. In a complex society like ours, there are certain matters on which the average person is judged to be incapable of making an informed choice. To outlaw marihuana smoking is not regarded as something like outlawing motorcycle riding, a taste or pursuit to which people have a right, however dangerous it may be to themselves and others. Instead it is regarded as something like outlawing the sale of a motorcycle part with a defective design. People might be willing to buy the motorcycle with that part only because of ignorance, carelessness, or impulsive folly. It has no value of its own, and the government has the right to protect people from it. Another example is the regulations insuring that electric chainsaws cannot be sold without shields to protect the user. Someone might want to buy a chainsaw without a guard, because it is cheaper or less clumsy to handle, but we do not respect that wish. Using a chainsaw without a shield may not be more dangerous than alpine climbing, but we also assume that the former has little value. These analogies could be specified further. Drugs with no accepted medical uses are treated like the defective motorcycles; they are pleasure vehicles that have a deadly flaw. Only in this case the flaw is integral to the design, so that the vehicle itself must be banned. Drugs with medical uses are more like the chainsaw— acceptable as a tool for certain purposes, but only with safeguards that cannot be left to the individual user to supply.

Certainly drugs are potentially dangerous instruments that can be ignorantly misused and sometimes produce serious ill effects. But it is not clear that the risk of, say, marihuana smoking is so obviously great, the benefit so obviously negligible, and the degree of ignorance on the part of the consumer so substantial that the drug has to be treated as something from which all consumers must be protected. Consumer protection laws imply that sometimes we cannot rely on the commonsense rule that greater dangers are also more obvious and therefore more likely to be avoided. In a case like the chainsaw, the danger might be too remote and contingent to outweigh the immediate inconvenience of a safety device if there were no special regulation. But presumably the more one thinks about it, the more one is likely to demand a chainsaw with a shield. It does not happen that way in the case of drugs. Studies show that the *less* people know about the actual effects of recreational drugs, the more dangerous they consider the drugs to be.[3] This has even been used as an argument against drug education in schools. If people who know more about the facts are wrong in their attitudes, which is quite possible, the mistake is obviously not caused by the kind of ignorance or preoccupation usually said to justify consumer protection laws.

Another peculiarity of drug laws as a form of consumer protection can be illustrated by comparing them with pollution control laws. Here we have to balance two goods, industrial productivity and a clean environment. We try to estimate how much loss in production we can tolerate for the sake of a given degree of improvement in air or water quality. If we followed the same policy in regulating drugs, we would try to estimate how much loss in the benefits of recreational drug use we should tolerate for the sake of a given reduction in their ill effects. But in fact the issue does not even arise, since benefits from recreational drug use are not conceded to exist for policy-making purposes.

A more striking comparison can be drawn with the laws on seatbelts in cars. It took a long political struggle to get them installed, and now most people will not use them. But there is not the slightest chance, in this country, that wearing seatbelts will be made compulsory. Most legislators who regard driving without a seatbelt as a right would reject as absurd the notion of a right to smoke marihuana. And yet the argument against individual freedom of choice seems at first glance much greater in the seatbelt case. The risk of driving without a seatbelt is overwhelmingly greater than any benefit, a fact not nearly so obvious in the case of marihuana; and driving without a

seatbelt is not an active taste or pursuit like marihuana smoking, so it is more likely to be done automatically, without conscious choice or genuine thought—the kind of situation in
which people need protection most.

Driving without a seatbelt resembles drug use because it
may cause accidental harm. Looking at television is sometimes
said to resemble drug use in another way: it is considered an
addictive, psychologically pernicious, and socially debilitating
practice, especially dangerous to children, which can seriously
damage the quality of life. But we have no laws defining a permissible amount of television viewing, just as we have no law
requiring the use of seatbelts. It is true that these matters are
not left to individual choice everywhere. In Norway, for example, drivers are fined for not wearing seatbelts; in some countries television hours are restricted by law. What is interesting
is that drug use, except for alcohol, is left to individual choice
nowhere. Drug laws are not typical protective laws, but a special set of regulations for a very special case.

Another anomaly makes this clearer. If it ever became necessary for the government to expend a vast amount of money
and manpower to curb an organized illicit traffic in chainsaws
without shields, or most other commodities forbidden by consumer protection laws, the law would probably be repealed. If
people wanted the commodities so much, we might conclude
that they have a legitimate interest and value strong enough to
outweight any argument for prohibition. In other words, we
would handle the problem as we handle mountain climbing,
hang gliding, or motorcycle racing; we would treat it as a matter of preferred tastes and activities (however questionable)
rather than consumer error. But it is considered normal and
necessary for the police to devote a large part of their resources
to combatting the drug traffic, and the size of the problem is
not regarded as a reason to change our way of dealing with it.

It is also significant that every activity connected with any
of the banned drugs is a crime, including simple possession. At
one time possession of some of these drugs was a felony, and
in some states sale of a few grams is still subject to the same
punishment as rape, armed robbery, and second-degree murder. Obviously more powerful passions are at work here than
those that produce the average consumer safety law.

If the comparison with seatbelts or chainsaws sounds
strained, that is because psychoactive drug control is usually

regarded as something more than an ordinary health and safety measure. Our drug laws are framed and justified under the Constitution as commercial regulations, consumer protection laws. The Pure Food and Drug Act of 1906, directed largely at opiates and cocaine, was one of the first modern federal consumer protection laws. But the regulations for pleasure drugs or drugs of abuse or "controlled substances" are not the same as the controls on other drugs. The two systems have different historical origins and remain different today, despite some fusion and overlap.

The distinction between prescription and over-the-counter drugs had no legal significance in the United States at the federal level until 1938. Before then, consumers had the right to choose most drugs for themselves, with or without the advice of doctors. Since the Food, Drug, and Cosmetic Act of 1938, choices about drugs in general have gradually been removed further and further from the consumer, transferred first to doctors and then to the government.[4] But the regulation of so-called narcotics started long before 1938. Under federal law, consumers have not had the right to choose opiates and cocaine since 1914, or marihuana since 1937. The government started telling doctors that using opiates to maintain an addict was not a legitimate medical practice long before it would have dared to substitute its judgment for a doctor's about any other kind of drug therapy. And the usual forms of medical control remain inadequate for these drugs: they require special criminal legislation (the Comprehensive Drug Abuse Prevention and Control Act of 1970) and a specialized federal agency (the Drug Enforcement Administration) with a budget of several hundred million dollars a year.

The controls on pleasure and performance-enhancing drugs are not just protection for consumers, but also a way of containing what is believed to be a threat to the social fabric and the moral order. This is the other analogy popularly used in thinking about drugs; they are a vice—part of the same group of problems that includes prostitution, pornography, and gambling. Iran since the revolution of 1979 supplies another example, more interesting because more alien. A police campaign has been instituted there against alcohol, drugs, and Western popular music, which are all regarded as parts of a single social problem. The music has to be outlawed because it is "addictive," causes disorderly conduct and sexual excess, and ultimately produces undesirable moral and social changes.

In other words, the Iranians think about music the way they and we think about drugs (and a few other vices). We do not believe that governments have a right to issue decrees on the forms and uses of music, and most of us do not regard musical listening habits as a moral issue. What makes us so sure that the Iranians are not right—that legal restraints on music are not just as desirable as legal restraints on drugs?

The Iranian notion of a moral order is much more comprehensive than the one prevalent in Western liberal societies. We are inclined to say that in its campaign against rock music, as in the similar campaign against homosexuality, Iran is legislating not morals but taste. It is a feature of liberal societies that more and more things once considered to be moral issues become matters of taste—or, as John Stuart Mill would have put it, tastes and pursuits. Besides, we do not endorse the idea of a universal moral order backed either by God's commandments or a natural law that prescribes the proper ends of humanity. Since our form of community is not based on fixed and shared moral assumptions, which are assumed to be universally valid, there is no easy way to distinguish between a breakdown in the social order and the emergence of a new moral consensus. We have no generally accepted code of sexual conduct or musical practice. When the Supreme Court declared contemporary community standards to be the basis for judging whether a legal definition of obscenity was constitutional, it satisfied no one, because people cannot agree even on what these standards are, much less on why they should have the authority of law. In these circumstances we have to fear that a consensus of prejudices and passions will present itself as the community's morality. What Mill called "the tyranny of prevailing opinion and feeling"—especially the latter—can easily disguise itself as a moral imperative to prevent social change.

As a result, in the West we are reluctant to justify legislation by an appeal to what for us have become floating generalities—the social fabric or the moral order. We try, at least, to be more precise about what we are protecting—such things as public health, public tranquility, esthetic values, productivity, and the welfare of children—and we are at least ostensibly committed to judging whether the practice we intend to legislate against is actually a danger to these interests.

Music, sex, and gambling evoke less concern than drugs. The greatest authoritarian philosopher, Plato, placed music and sex in the domain of government regulation; the ruling Iranian clerics and other successors in the authoritarian tradition tend to agree. Liberal theory, and now also liberal society, have put this in doubt. But all of us except a few libertarian eccentrics think that drug use requires government restraint. A police campaign against popular music looks absurd to most of us, and a government attack on homosexual behavior looks sinister to some of us; but if there is any point of view from which a war on drugs looks absurd or sinister, it is nowhere to be encountered today.

The familiar explanation is that drugs are not like sex and music: they are poisons that can produce deadly habits as well as accidents and crimes, a threat to life and health. They are not outlawed just to preserve some vaguely defined community bonds, and the objection to them is certainly not a matter of taste. But why then are the drug laws so severe compared to those controlling other dangerous commodities and activities? Why do we make an exception for alcohol, one of the most dangerous drugs? Why do Iranians blame rock music and drugs for some of the same evils? These are the questions that made it necessary to raise the issue of legislated morality in the first place.

Whether or not they are promulgated in the name of morality, laws meant to protect the social fabric or community bonds, and laws with more precise aims, such as promoting productivity or public tranquility, are vulnerable to the objection that it is very hard to adjust means to ends when the harm contemplated is so vague and all-encompassing. In discussing the loss of community or the decline of productivity, it is almost impossible to distinguish causes from symptoms. We cannot tell whether the troubling behavior—say, gambling, deviant sex, prostitution, or pornography—is the source of any social problems that seem to be associated with it, or an ineffectual attempt to cope with them, or neither.

So we become confused and uncertain when we try to justify laws against such behavior. Unlike most consumer protection laws, these laws apparently lack a clear, limited purpose and well-defined effects.

Drugs are different. Cause and effect seem clear enough here. A simple chemical is put into the body and makes people sick or dangerous. Here talk about poisoning is more than a metaphor; the consumer protection analogy is available to fall back on. But we have already pointed out that if drug control were merely a matter of consumer safety, the unusual severity of the laws and the enormous enforcement problem would be inexplicable; so would the peculiar fact that well-informed people tend to see less danger in drugs than ill-informed people. Drugs are a dangerous commodity, but they are also the source of a disapproved way of living, so they are subjected to two kinds of moral and legal censure. But the rules appropriate for regulating vice (or ensuring public order) and those appropriate for regulating consumer safety are normally of very different types. The problem is circumvented by using a broad conception of public health.

In its 1980 campaign platform, the Republican party called the American drug problem a "murderous epidemic." Many social problems are described from time to time as epidemic. This is a legitimate sense of the word, but in its primary use it still refers to physical disease. It is significant that drug use evokes the image more often and more powerfully than any other social problem. Traditionally, a plague was the vengeance of the gods on a community whose leaders had transgressed the moral law: the plagues of Egypt in the Old Testament or the plague of Thebes in Sophocles' *Oedipus*. Epidemic physical disease can no longer be regarded as a sign of moral disorder because we know its causes, but drug abuse is another matter. By calling it an epidemic, we suggest a public health campaign that is also a moral crusade.

If drug abuse is a communicable disease, and drugs are a menace like the typhoid bacillus or the smallpox virus, the reasons for intervention become overwhelming. The imagery of disease has tremendous social potency. It eliminates most moral and political doubts, since disease has nothing to do with free action. Preventing an epidemic of typhoid presents no moral problems, so why should preventing an epidemic of drug abuse create any? The infectious disease carrier, the Typhoid Mary, has to be quarantined, so why shouldn't heroin addicts be locked up? If the persuasion and imitation by which drug use spreads are regarded as a form of infection, the drug as a disease agent, and the drug user as a carrier, freedom and individual desires obviously deserve no consideration. There is no such thing as freedom to be infected with typhoid. The thought that everyone might perform homosexual acts or read pornography may be troubling, but on this analogy the prospect that everyone will constantly use drugs for pleasure is worse—as though everyone were to come down with cholera. We no longer think of people doing things to themselves that others consider bad for them; instead we picture an external agent, the drug, invading the individual and social body and spreading irresistibly. Preventing disease has been regarded as a government responsibility for a hundred years; that is the meaning of public health medicine. And in this situation we have to act even on remote and indirect possibilities of harm, as we do when we quarantine and vaccinate.

Above all, the language of disease dismisses the question of whether anyone ever gets any good out of using a drug for pleasure, and it denies any important resemblance between drug use and other tastes, pursuits, and indulgences. We know that it would be absurd to speak of a murderous epidemic of mountain climbing. If we can solemnly refer to a murderous epidemic of drug use, we do not have to consider whether, say, marihuana or LSD causes as many deaths as mountain climbing. It also becomes harder to think of marihuana use as something like an eating habit—the sort of indulgence that may be good, bad, or morally indifferent depending on the persons, amounts, and circumstances involved, and that is normally subject only to informal social pressure and persuasion. A community threatened by infectious disease must take the necessary measures to stop it.

The vocabulary of public health medicine also permits a smooth transition from physical health to psychological and moral health and finally to social health: the "murderous epidemic" is crime and illness at once, without careful distinction. In this way consumer safety becomes mixed with morality; the two different kinds of justification reinforce each other. Mill rejected alcohol and other drug control laws on the ground that they gave the majority a legally enforceable interest in every citizen's moral, intellectual, and physical perfection, to be defined as the majority saw fit. A plausible reply is that curbing mass drunkenness is not a form of officious meddling aimed at remaking some people in the image of others; it is a matter of elementary public health and social order. Like the idea of social order, the idea of health is ambiguous; it has biological, social, and moral elements. The World Health Organization once defined it as not just freedom from disease, but total well-being—physical, mental, and social. By this definition, anything desirable is healthful and vice versa—an open-ended view of health characteristic of what has been called the therapeutic state, a twentieth-century development that might have shocked Mill. Alcohol control laws have not changed much, but in the control of other drugs another point of view has prevailed.

Besides, disease is the realm of medical expertise; "drug abuse" is often simply identified with nonmedical use. In the twentieth century, medicine has finally become a science, so medical control has come to imply objective standards and an unchallengeable authority; the ordinary citizen has no more claim to judge the proper uses of drugs than the theories of biology. Mill did not approve of the taste for opium, but he objected to laws making it impossible to indulge the taste without a doctor's prescription. Today, however, taking opiates is no more considered a matter of taste than using an X-ray machine. To do it without supervision is simply a dangerous misuse of an instrument valuable only for certain purposes understood by the medical profession. And the rules against taking risks in medicine are much stricter than the rules against taking risks in the pursuit of pleasure or ambition. If we had developed a biochemistry of hang gliding, and the sport had somehow come under medical supervision, it would have been banned long ago.

With certain familiar exceptions, then, chemicals that people want to ingest for any purpose except food are treated as resembling plague germs, unless they are used under more or less the same restrictions as a medicine like pencillin. The language of medicine was called on in the first place to make the special severity of drug controls plausible where justifications based on simple consumer protection or legal moralism seemed inadequate. But since the usual forms of medical control are obviously too mild, we continually turn back to other justifications to supplement the medical one. This circle is never made explicit, since the transitions are blurred by the ambiguous meaning given to terms like "public health." Since this special public health problem includes psychological, social, and moral health, the use of drugs for pleasure is believed to present three kinds of threat to human welfare. It is an offense to morals or a danger to the social fabric; in some ways it is also like an epidemic disease; and in some ways it resembles the ignorant use of a dangerous instrument like a chainsaw. Social attitudes and legal regulations conform to each of these three analogies in different ways, and each model in turn reinforces the others at their weak points to supply reasons for stricter controls. At all times we continue to reject the idea of treating drug use as a risky taste or pursuit that nevertheless may have some value for some people, like climbing mountains, riding motorcycles, or eating rich food. Marihuana smoking can be seriously described as an epidemic, but mountain climbing as "getting high" is only a joke.

And yet we know that it is not absurd to treat drug use as a matter of taste, because that is the way we have always regarded alcohol. It is not just that alcohol is legal and most other pleasure drugs illegal. Our whole public discourse about alcohol, even that of prohibitionists, is different from the way we talk and think about other drugs. No one pretends that alcohol prohibition can be treated as a consumer safety measure. For example, during the campaign for Prohibition, it was finally understood that a ban on alcohol would require a constitutional amendment, unlike other drug laws and consumer laws in general. To begin outlawing opiates and cocaine, all we needed was a law based on the taxing power granted to Congress in the Constitution (the Harrison Act of 1914 was formulated as a revenue measure, because most police powers were thought to be reserved to the states by the Tenth Amendment). But in 1915 the Supreme Court declared unconstitutional a Kentucky state law banning the possession of alcohol. Possession of alcohol was never a crime even under the Volstead Act, and since 1933 we have relied almost entirely on appeals to self-restraint in dealing with alcohol problems—a limitation unimaginable in the case of any other drug. Alcohol has never been under medical control, either, except briefly during Prohibition. We do use disease analogies in talking about alcohol abuse, but the emphasis is on the susceptible individual and not the irresistible infectious agent. Few people, except in Moslem countries, dream of an alcohol-free utopia, and even fewer want to impose it by law. Alcohol escapes the net of analogies used to control other drugs.

The most common argument used to justify this inconsistency is that alcohol prohibition, unlike other drug prohibitions, does not work. It is true that, even in a completely free market, alcohol might be the most popular drug; that makes banning it harder. But prohibition probably does reduce the consumption of alcohol and the harm done by drinking: American alcohol use declined in the 1920s, and the incidence of diseases like cirrhosis of the liver dropped sharply.[5] Prohibition is also said to cause more harm than it prevents, but if that is so it should be just as true of other drug laws. Alcohol prohibition in the twenties and marihuana prohibition today have produced the same nasty side effects: costs of arrest and punishment, growing disrespect for the law, organized criminal violence, police corruption and oppression, poisonous adulteration, and misrepresentation. The parallels are ridiculously precise, down to the tendency for research to be monopolized by prohibitionists and the national commissions (the Wickersham and Shafer Commissions) which were appointed to investigate the problem and came to self-contradictory, but cautious and therefore politically acceptable, conclusions. In fact, allowing for the difference in the size of the original problem, alcohol prohibition probably worked just about as well (or badly) as present drug prohibition laws work. Repeal came not because prohibition was totally ineffective, but because we decided—though we seldom put it to ourselves this way—that we valued the pleasure of convenient legal alcohol more than we feared an increase in drunkenness and alcoholism. It is still unthinkable to make the same kind of balancing judgment about any other drug, even to come to a different conclusion.

We concede that alcohol is a very dangerous substance and creates a vast health problem, but can also be a harmless indulgence. But apparently the strain of tolerating the ambiguity is great, because—at least officially—we are unable to do the same for any other drug, even when there is little evidence that the drug could ever be as dangerous as alcohol. Unofficially, the situation is different. In public discussions, drugs are a dangerous commodity, a public health problem, and a moral menace that requires the full force of the law; but in polls a large part of the population says that drug use should be a matter of individual choice.[6]

There is probably as much honest confusion as hypocrisy in all this. In modern society experiences and behavior that are hard to classify make us anxious. Drug use for work or pleasure is both a cement for the social order and a threat to it; the drug user is free and yet often, somehow, compelled; the drugs are medicine, but also vice or even a kind of infectious disease. The existing system of drug control is entirely a product of our century both institutionally and conceptually. Facing an ambiguous situation and feeling a new impulse toward rationalization and control, we have allayed our anxiety by fixing different le-

gal and social categories for different drugs. If they seem hard
to justify in the abstract, perhaps they represent some collect-
ive historical wisdom. Certainly no great changes are likely.
But at least we will be able to use these categories more wisely
if we remember what was left out in formulating them.

We should acknowledge our own confusion and not cover
it up with self-righteous rhetoric. We can afford more confidence in the judgment of the average person and the recupera-
tive powers of the community; if any drug could destroy the
fabric of society, alcohol would have done it by now. Sociolo-
gists have been quick to write condescendingly of a "symbolic
crusade" and "moral entrepreneurship" in the campaign for
prohibition of alcohol, but this kind of irony comes less easily
when other drugs are involved. The notion of preserving a
moral order is usually a poor basis for policy, if only because it
tends to be applied mainly in areas where true cause and effect
are hard to determine. We should concentrate on health and
safety hazards of which drug use is clearly a cause and not a
symptom. The imagery of epidemics should also be restrained.
Once we become more conscious of the way we are using
analogies, we may begin to question the wisdom of subjecting
drug use to harsher legal restraints than other forms of
risk-taking.

References

[1] Hugh Cann Phallow (A. S. Carlin), "Alpinism: Social, Scientific, and Treat-
    ment Aspects of Getting High and its Prohibition," *Journal of
    Psychedelic Drugs* **10**, 1978, 77–78.
[2] J. R. Torg, R. Truex, T. C. Quedenfeld, A. Burstein, A Spealman, and C.
    Nicholas, "The National Football Head and Neck Injury Registry,"
    Journal of the American Medical Association **241**, 1979, 1477–1479.
[3] D. Glaser and M. Snow, *Public Knowledge and Attitudes on Drug Abuse in New
    York State*, New York State Narcotics Control Commission, 1969; also
    see J. D. Swisher, R. W. Warner, and E. L. Herr, "Experimental
    Comparison of Four Approaches to Drug Abuse Prevention among
    Ninth and Eleventh Graders," *Journal of Counseling Psychology* **19**,
    1972, 331–332.
[4] Peter Temin, *Taking Your Medicine: Drug Regulation in the United States*,
    Cambridge, MA, Harvard University Press, 1980.
[5] J. C. Burnham, "New Perspectives on the Prohibition 'Experiment' of the
    1920's," *Journal of Social History* **2**, 1968–69, 51–68.
[6] Glaser and Snow, op. cit.

The Social Dilemma of the Development of a Policy on Intoxicant Use

Norman E. Zinberg

It is generally believed that in the area of drug use only two types of behavior are possible—abstinence, or unchecked excess leading to addiction. Despite massive evidence to the contrary, many people remain unshaken in this belief. The conviction persists that because of their pharmacological properties, the psychedelics, heroin, and, to a lesser extent, marihuana cannot be taken on a long-term, regular basis without causing serious problems. Until after 1970, even scientific researchers were strongly influenced by this moralistic view that any kind of illicit drug use was bad and inevitably harmful—"addictive"—and that abstention was the only alternative.[1] The unfortunate condition of heroin addicts and other compulsive users was regularly cited as "proof" of this "pharmaco-mythology."[2] Related to that view was the equally strong conviction that individuals who sought out such drugs were almost always suffering from profound personality disorders.

Studies of drug consumption therefore tended to equate use (any use) with abuse, and they seldom regarded occasional and moderate use as a viable pattern.[3] If the possibility of nonabusive use was acknowledged, it was usually treated as a very brief transitional stage leading either to abstinence or to some sort of trouble, probably addiction. Research emphasis was placed primarily upon determining the potentially harmful effects of illicit drugs, and secondarily on the study of the personality disorders resulting from such use (which at the same time were assumed to be responsible for that use).

The New Perspective on Control

Even before 1970, however, it was recognized that in order to understand how control of a substance taken into the body was developed, maintained, or lost, different patterns of consumption had to be compared. During the past decade this principle, which had been successfully applied to the comparative study of patterns of alcohol use—alcoholism as against social or moderate drinking—began to be applied to the use of such illicit drugs as marihuana, LSD, cocaine, and heroin, with the result that a wide range of using patterns is now recognized.[4]

This recent interest in the comparative study of drug use and abuse is attributable to two factors. First, despite the vast increase in the number of marihuana users, the widely held expectation of health hazards has not been fulfilled.[5] In addition, most of the marihuana use has been occasional and moderate, rather than intensive and chronic.[6,7] These discoveries have spurred public and professional recognition of the fact that moderate use of illicit substances may be possible and that the question of how control is achieved at various levels of consumption deserves research attention. The second factor is the pioneering work of a few students of heroin use, who have dared to follow the logic of their own findings even when these have deviated from the mainstream view. The most influential work has been that of Lee N. Robins, who studied drug use among Vietnam veterans (to be discussed later). She found that consumption of heroin did not always lead to addiction or dysfunctional use, and that even when addiction occurred it was far more reversible than was popularly believed.[8]

As the view moderated that illicit drugs were in a class by themselves, comparisons began to be made with the licit drugs and other substances. It was found that many of these—tobacco, caffeine, sugar, and various food additives—are potentially hazardous to health. Other research demonstrated that even with prescribed drugs, "good" drugs for "good" purposes, the failure to use them in the way the physician intended can also be hazardous and may in itself constitute a major public health problem. It seemed that just as the mythology that illicit drugs are altogether harmful was giving way, so too was the mythology that all licit substances are benign.[9] The result has been a new interest in studying the commonalities in the ways of controlling the use of a wide variety of substances, both licit and illicit.

The purpose of the project that I and my colleague Richard C. Jacobson (succeeded in 1974 by Wayne M. Harding) undertook in 1973 with the support of The Drug Abuse Council was to study the patterns of controlled (moderate, occasional) use of illicit drugs. We chose marihuana, the psychedelics, and the opiates because they were the drugs most widely used at that time. In 1973 formal efforts had just begun to decriminalize the private use of marihuana; psychedelic drug use was increasing at a very high rate (131 percent, according to the National Commission on Marihuana and Drug Abuse)[10]; and the concern about a heroin "epidemic" was declining. Cocaine had not yet become popular, nor had publicity been given to PCP. The two underlying hypotheses of our project were far more controversial in 1971–1972, when they were first conceived, than they would be today, although even now they have not been generally accepted.

The leading hypothesis was that in order to understand what impels someone to use an illicit drug and what its effects on the user are, one must consider three determinants: *drug* (the pharmacologic action of the substance itself), *set* (the attitude of the person at the time of use, including his personality structure), and *setting* (the influence of the physical and social setting within which the use occurs).[11,12] Of these three, setting had been the least considered and understood and therefore it was to be an important focus of our investigation, although the basic emphasis was always to be on the interaction of the three variables.[13,14]

The other hypothesis (which will be discussed here first) concerned the manner in which the "setting" variable functioned to exert some control over drug choice, level of use, and effect. The project team planned to study the ways in which users developed informal social controls, that is, socially determined sanctions (or rules) for use and social rituals or patterns of behavior. Although this form of social learning about intoxicants was not new, we needed to understand the conditions under which such informal social controls could (or could not) be disseminated and whether they could be bolstered by formal institutional practices (laws, enforcement procedures) and by other social controls.

Social sanctions define whether and how a particular drug should be used. They may be either informal and shared by a group, such as the maxims applied to alcohol use, "Know your limit" and "Don't drive when you're drunk," or formal, such as the laws and governmental policies aimed at regulating drug use.[15,16] Social rituals are the stylized, prescribed behavior patterns surrounding the use of a drug; they may apply to the methods of procuring and administering the drug, the selection of the physical and social setting for use, the activities undertaken after the drug has been administered, and the ways of preventing untoward drug effects. Rituals thus serve to buttress, reinforce, and symbolize the sanctions. In the case of alcohol, for example, the statement, "Let's have a drink," automatically exerts control by using the singular term "a drink."

Informal social controls apply to all drugs, not just to alcohol, and operate in a variety of social contexts, ranging all the way from very large social groups, representative of the culture as a whole, down to small, discrete groups.[17] Certain types of special-occasion use involving large groups of people—beer at ball games, drugs at rock concerts, wine with meals, cocktails at six—have become so generally accepted that few if any legal strictures are applied even if such uses technically break the law. And if the culture as a whole thoroughly inculcates a widespread social ritual, it may eventually be written into law, just as the socially developed mechanism of the morning coffee break has been legally incorporated into union contracts. The TGIF (Thank God It's Friday) drink may not be far from acquiring a similar status. For small groups, however, the sanctions and rituals tend to be more diverse and more closely related to circumstances. Nevertheless, some caveats may be just as firmly upheld, such as: "Never smoke marihuana until after the children are asleep"; "Only drink on weekends"; "Don't shoot up until the last person has arrived and the doors are locked."

The existence of social sanctions and rituals does not necessarily mean that they will be effective, nor does it mean that all sanctions and rituals were devised as mechanisms to aid control. Of more practical importance, however, than the purpose of the sanction is the way in which the user handles the conflicts between the different sanctions. For illicit drugs, the most obvious conflict is that between formal and informal social controls, that is, between the laws against use and the social group's condoning of use. The teenager attending a rock concert is often pressured into trying marihuana by his peers, who insist that smoking is acceptable at that particular time and place and will enhance his musical enjoyment. The push to use may include a control device, such as, "Since Joey has a cold and isn't going to smoke, he can drive," thereby honoring the "Don't drive after smoking" sanction. Nevertheless, the decision to use, so rationally presented, conflicts with the law, causing concern that the police may intervene. Such anxiety interferes with control.[18] In order to deal with the conflict, the user will probably express more bravado, paranoia, or antisocial feeling than if he had patronized one of the little bars set up near the concert area to sell alcohol. It is this kind of mental conflict that makes control of illicit drugs more complex and difficult than the control of licit drugs.

Although the existence of social controls, particularly in the case of illicit drugs, does not always bring about moderate use, it is the reigning cultural belief that use should always be moderate and behavior correct. This seems to be the chief reason why the power of the social setting to regulate intoxicant use effectively has not been more fully recognized. Controlled use has been considered so patently "right" that understanding how control is achieved seems unnecessary. The idea that users of intoxicants should always behave properly stems from the moralistic attitudes pervading our culture, attitudes that are almost as marked in the case of licit drugs as in the case of illicit drugs. Yet on some occasions—at a wedding celebration or during an adolescent's first experiment with drunkenness—less controlled behavior is culturally acceptable. Though we should never condone the excessive use of an intoxicant, it has been recognized that when such limit-testing occurs, it does not necessarily signify a breakdown of overall control.

Social Learning

Social learning about drug use occurs when young people internalize the informal social controls—sanctions and rituals—

that relate to the use of a particular drug. In most sectors of our society, informal education in the use of alcohol is readily available. Few children grow up without an awareness of a wide range of behaviors associated with alcohol use, learned from television, movies, the print media, by observing family and family friends, and often by taking a sip or watered-down taste of a grownup's drink. Later, in a peer group, they may begin to drink and even, as a rite de passage, to overdo it, but they usually know what they are doing and what the sanctions are. Social learning about illicit drugs is far less straightforward. Since the family cannot legally use such drugs and inculcate the informal social controls that govern moderate use, most learning goes on underground, where youngsters try out the new drug with their peers. Nevertheless, each group of users develops its own sanctions and rituals and teaches them to newcomers.

Attempts made in the late 1960s and early 1970s to transform this informal learning process into formal drug education courses, chiefly intended to discourage any drug use, have failed.[19] Paradoxically, they have actually stimulated drug use on the part of many young people who were previously uncommitted and at the same time have confirmed the fears of those who were already excessively concerned. This reverse of what was sought was generally recognized at a meeting of the US Coordinating Council on Drug Education in 1964, leading to a declared moratorium on drug education films and classroom materials.[20-24] The materials and films generally were heavy-handed and prohibitionistic even when billed as exercises in decision-making rather than outright antidrug education. These findings were confirmed in a study by Yankelovich, Skelly and White, Inc. for The Drug Abuse Council in 1975.[25]

One might ask whether it is possible for formal education to codify social sanctions and rituals in a reasonable way for those who have somehow been bypassed by the informal process, since the cultural moralism pervading such courses precludes the possibility of discussing the informal social controls that seem to condone use. This question cannot be answered until our culture has accepted the use not only of alcohol, but possibly of other intoxicants so that teachers may be free to explain how these substances can be used safely and well. Teaching safe use is not intended to encourage use. Its purpose is the prevention of abuse, as is the case with the few good sex education courses in existence today whose primary aim is to teach the avoidance of unwanted pregnancy and venereal disease, rather than to either encourage or discourage sexual activity per se.

Whatever may happen to formal education in these areas, the natural process of informal social learning will continue. The power of this process is illustrated by two relatively recent social events: the use of psychedelics in the United States in the 1960s and the use of heroin during the Vietnam War.

First, the psychedelics. Following the Timothy Leary "Tune In, Turn On, and Drop Out" slogan of 1963, psychedelic use launched the so-called drug revolution and became a subject of national hysteria. The use of these drugs, known then as psychotomimetics (imitators of psychosis), was widely believed to lead to psychosis, suicide, or even murder.[26,27] Equally well publicized was the contention that they could bring about spiritual rebirth, mystical oneness with the universe, and other metaphysical experiences.[28,29] Certainly there were numerous cases of not merely transient but prolonged psychoses following the use of psychedelics. In the mid-sixties psychiatric hospitals like the Massachusetts Mental Health Center and Bellevue in New York were reporting that as many as one-third of their admissions resulted from the ingestion of these drugs.[30] By the late sixties, however, the rate of such admissions had declined dramatically. At first, observers concluded that this decline was the result of fear tactics: the warnings about the various health hazards, the chromosome breaks and birth defects, that were reported in the newspapers—and that later proved to be false. Eventually, although psychedelic use continued to be the fastest growing drug use in America up to the end of 1973, the dysfunctional sequelae virtually disappeared.[31] What then had changed?

Researchers found that neither the drug itself nor the personality of the user was the most prominent factor in those painful cases of the sixties. The retrospective McGlothlin and Arnold study[32] revealed that before the early sixties responses to these drugs had included none of the horrible experiences that occurred in the mid-sixties. Another study, described in the book "LSD: Personality and Experience,"[33] investigated the influence of personality on psychedelic drug experience before the so-called drug revolution. It found typologies of response to the drugs, but no one-to-one relationship between untoward reaction and emotional disturbance. And Howard S.

Becker in his prophetic article of 1967 compared the then current anxiety about psychedelics with the anxiety about marihuana in the late 1920s, when several psychoses had been reported.[34] Becker hypothesized that the psychoses came not from the drug reactions themselves, but from the secondary anxiety generated by unfamiliarity with the drug's effects and intensified by media publicity. He suggested that the eventual disappearance of such unpleasant reactions had taken place as familiarity with the effects of marihuana had increased, and he correctly predicted that the same thing would happen with the psychedelics.

The power of social learning also brought about a gradual change in the reactions of those who expected to gain insight and enlightenment from the use of psychedelics. Interviews have shown that the user of the early 1960s, with either his great hopes of heaven or his fears of hell, as well as his lack of any sense of what to expect, had a far more extreme experience than the user of the 1970s, who had been exposed to a decade of interest in psychedelic colors, music, and sensations. The later user, having been thoroughly prepared, even though unconsciously, for the experience and accordingly responding in a moderate way, might simply remark, "Oh, so that is what a psychedelic color looks like," instead of undergoing a "mystical" transformation.

The second example of the influence of the social setting and of social learning in determining the consequences of drug use comes from Vietnam. Estimates noted by Robins and her colleagues in 1977[35] indicated that at least 35 percent of the enlisted men (EMs) used heroin and of these 54 percent became addicted. Statistics from those US Public Health Service Hospitals engaged in detoxifying and treating addicts showed a recidivism rate of 97 percent, but some observers thought it was even higher. Once the extent of the use of heroin in Vietnam became apparent, the great fear of Army and government officials was that the maxim, "Once an addict, always an addict," would operate; and most of the experts agreed that this fear was entirely justified. Treatment and rehabilitation centers were set up in Vietnam, and the Army broadcast its claim that heroin addiction stopped "at the shore of the South China Sea." Yet, as virtually all professional observers agree, those programs were total failures. Often people used more heroin in the rehabilitation centers than they had while on active duty.[36]

Nevertheless, as the study by Robins and her colleagues has shown,[37] most addiction did indeed stop at the South China Sea. For those addicts who had left Vietnam and gone home to the United States, recidivism was under 10 percent, very much lower than the figure given for treated addicts who had stayed on in Vietnam. Apparently it was the abhorrent social setting of Vietnam that led men who ordinarily would not have considered using heroin to use it and often to become addicted to it. Evidently they associated its use only with Vietnam, much as hospital patients who are receiving large amounts of opiates for a painful medical condition associate the drug with the condition and usually do not crave it after they have left the hospital.[38-40]

To return to the first example—psychedelic drug use in the 1960s—it is evident that control over use of those drugs was established by the development in the counterculture of social sanctions and rituals very like those surrounding alcohol use in the culture at large. "Only use the first time with a guru" was a sanction or rule that showed neophytes the importance of using the drug initially with an experienced user who could reduce their secondary anxiety about what was happening. "Only use at a good time, in a good place, with good people" gave sound advice to those taking a drug that would sensitize them intensely to their inner and outer surroundings. In addition, this sanction conveyed the message that the drug experience could be just a pleasant consciousness change instead of a taste of heaven or hell. The specific rituals that developed to express these sanctions—when it was best to take the drug, how, with whom, and what was the best way to come down— varied from group to group, though some rituals spread from one group to another.

It is more difficult to document the development of social sanctions and rituals in Vietnam. Most of the early evidence indicated that heroin was used heavily in order to obscure the actualities of the war, with little thought of control. Yet later studies showed that many EMs used heroin in Vietnam without becoming addicted.[41] More important, although 95 percent of all heroin-addicted Vietnam returnees, not only the original group, did not become addicted in the United States, a higher percentage did take heroin occasionally, indicating that they developed some capacity to use the drug in a controlled way.[42] Some rudimentary rituals, however, do seem to have been fol-

lowed by the men who used heroin in Vietnam. The act of gen-
tly rolling the tobacco out of an ordinary cigaret, tamping the
fine white powder into the opening, and then replacing a little
tobacco to hold the powder in before lighting up the OJ (opium
joint) seemed to be followed all over the country even though
the units in the North and the Highlands had no direct contact
with those in the Delta.[43] To what extent this ritual aided control is, of course, impossible to determine. But having observed
it many times, I can say that it was almost always done in a
group and thus formed part of the social experience of heroin
use. While one person was performing the rituals, the others
sat quietly and watched in anticipation. It would be my guess
that the degree of socialization achieved through this ritual
could have had important implications for control.

One other factor relating to social learning deserves mention. It is not generally recognized that the rise in interest in
consciousness-changing illicit drugs has closely followed the
rise in interest in consciousness-changing licit drugs. Until the
fifties, medical men had few psychoaffective substances to prescribe; those they had, such as the barbiturates or the amphetamines, had very general depressant and stimulant effects and
were not particularly specific. Until the late fifties, however,
tons of phenobarbital were being prescribed each year in the
hope that the drug would reduce anxiety for some people. Suddenly there was a technological revolution, and pharmacolo-
gists ingeniously developed class after class of new drugs. First
came the phenothiazines, with antipsychotic properties; then
the monoamine oxidase inhibitors for anxiety and depression;
then the tricyclic antidepressants and the benzodiazapines
with antianxiety properties appeared. Each one of these new
classes of antipsychotic, antidepressant, or antianxiety medications spawned numerous variants and combinations intended
to achieve specific psychoactive changes. It would not be going
too far to say that the entire community mental health movement and the press for deinstitutionalization of the enormous
state hospital systems would not exist were it not for the
phenothiazines.

Not long ago I appeared on a panel with a prominent
psychopharmacologist who was describing the successful treatment of a complex case of mixed paranoia, depression, and
anxiety. The doctor described how, by carefully titrating a mixture of a phenothiazine, a tricyclic, and a benzodiazapine, he

had helped the patient significantly. In reply I said I had just
returned from Washington where young men and women in
important political posts had been telling me about their daily
lives: long hours of hard, intense work followed by a round of
cocktail parties. Not only did they enjoy the tension-relaxing
drinks, but the parties, attended by many Washington-based
power brokers, afforded them important opportunities to get
business done. In order to cut through the alcohol and get back
to work after the parties, they found that a little sniff of white
powder was just the ticket. Then, when they finally returned
home exhausted, puffing on a "joint" often helped them relax.
Was this careful titration of illicit drugs, I asked, really very different, except in its degree of emotional difficulty, from the
medication regimen described by the psychopharmacologist?

The unanticipated social consequences of technological
change are always hard to spot and to evaluate. The process of
social learning is by its nature inchoate and uncodified. In contrast to the technological innovations, the social consequences
take place slowly, often imperceptibly, and generally they can
only be identified in retrospect.

Controlled Use Versus Prohibition

People who either do not use intoxicants at all or who only
use them occasionally in order to be sociable find it hard to understand those who use intoxicants regularly because they enjoy the relaxation and sense of ease that come with being high.
This lack of understanding does not necessarily mean disapproval. In the case of alcohol, those who have that extra drink
or two at a cocktail party and lose a little control are accepted,
and particularly within the middle class, are looked after. In
other social groups permission to turn on with marihuana, to
take a sniff of white powder, or to tell of an experience with a
psychedelic may also be granted. Yet, because of the general
lack of understanding, disapproval of drug use, coupled with a
sense of moral outrage, is often expressed in our society. Moreover, the prohibition mentality—the determination to stop
whatever it is that appears offensive—may easily override the
desire of professionals and some users to find out how the safe
use of intoxicants can be achieved. For an interest in control
presupposes a wish to save what is of value in the drug experi-

ence while at the same time minimizing the possibilities of misuse. Also, and this point is far more controversial and is seldom
stated bluntly, an interest in controlling use instead of prohibiting it (which is our current social policy toward alcohol)
means accepting a certain level of problems resulting from that
use. Obviously, as with automobile accidents, a continuing effort must be made to reduce that level—although the current
controversies over automobile safety measures indicate that diverse economic, psychological, and political interests may interfere with this goal. In the case of automobiles, there is an
unspoken but universal acknowledgment that their use is so
essential that we must accept a casualty rate running to 55,000
deaths and 2 million injuries a year resulting from automobile
accidents.[44] Until there is some general appreciation of the
value of intoxicants to many individuals, there can be little possibility of serious debate over what is an "acceptable" casualty
rate for drug use. This situation poses at least two problems for
contemporary America.

First, although those interested in moderate intoxicant use
(and it is my impression that they make up a considerable majority of the population),[45] care a great deal about such use,
our society refuses to admit the importance of this concern or to
formulate coherent personal or institutional policies regarding
the use of intoxicants. Freud was undoubtedly right when he
said that the two most important things in life were "to work
and to love." Religion was once regarded as the third most important thing, but with the present decline of religious concerns, the consumption of food and intoxicants may be taking
third place in this hierarchy of interests. One striking difference, however, between the first two interests and the third
one is that whereas most people approve of work and love,
they seem, in this culture at least, to disapprove of an interest
in food and drink.

Second, although the United States has been in the throes
of a drug revolution since about 1963, with millions and millions of people—according to the National Institute on Drug
Abuse 1980 figures, 51 million users of marihuana, 10–12 million users of psychedelics, 15–20 million users of cocaine—
trying substances that previously had been used by only a very
small minority who were regarded as deviant, our society has
not yet awakened to the implications of the acceptance and use
of these substances. The drug revolution that began with the

increased use of psychedelics expanded during the middle and late sixties to include marihuana. Then in 1969 the heroin "epidemic" appeared, and the increased use of that drug preoccupied public attention through 1971–1972, leading to the creation of the Special Action Office for Drug Abuse Prevention (SAODAP) in the White House. Next, the National Commission on Marihuana and Drug Abuse (the Shafer Commission) was appointed, and it issued two important reports— *Marihuana: A Signal of Misunderstanding*[46] and *Drug Use in America: Problem in Perspective*.[47] Both of these reports, produced by a panel of distinguished and conservative (antidrug) professionals from various fields, expressed great concern about the difficulties that could arise from the extensive use of these newfound substances. Nevertheless, the commission's main message was a plea for society to find ways to come to peaceable grips with a phenomenon, the use of drugs, that was not going to go away. Although the reports went largely unheeded, they are noteworthy not only because their antihysteria message is central to the theme of this paper, but because they pay only slight attention to the use of cocaine, which since their publication has become the fourth most popular drug in America, after nicotine, alcohol, and marihuana, in that order.

It looks as though American society is carrying on a vast experiment aimed at finding out about various intoxicants and how each one can best be used. If such a social experiment is going on, surely an informed debate over what social cost society is willing to pay for the institutionalization of use of each specific intoxicant versus the social cost of efforts at prohibition is essential. By now it has become clear that our present prohibition policy is being maintained at a high cost. Yet debates over drug use continue to focus on the prohibition of all intoxicants as a class (alcohol excluded), rather than on how to deal with each drug separately.

Social Policy

Current social policy designates all drug users as criminals, deviants, or even "miscreants," and in response to that policy the medical establishment designates all users as mentally disturbed. These attitudes prevail despite the plea, made in several government reports during the 1970s, that a distinction be made between experimental, recreational use on the one hand and dysfunctional, compulsive use on the other. The Shafer Commission was the first to launch an eloquent plea for those distinctions. These two documents, intended by then President Nixon (who appointed the members of the Commission) as plans for the elimination of drug abuse, called for more distinctions to be drawn among users. The *White Paper on Drug Abuse*[48] also acknowledged that the goal of eliminating "drug abuse" from our society is unrealistic, and the *Federal Strategy for Drug Abuse and Traffic Prevention*[49] announced that "drugs are dangerous to different degrees." In 1978 the *Report to the President* from The President's Commission on Mental Health[50] concluded that not only was it important to make distinctions, but that not to make them would be extremely costly to society, much as the Volstead Act had been costly. Among those costs were loss of respect for the law in general, since the drug laws were being persistently flouted; increased corruption among enforcement and other public officials; and a virtual consensus among informed citizens that although they might support the drug laws in principle, they would try to circumvent them if relatives or friends were involved. It is also clear that the labeling of people as criminal who would otherwise probably not be considered so has been greater under the drug laws than under the Volstead Act. The large majority of those affected by these laws are young, the penalties have been more severe and therefore more life-changing, and often the only choice has been between accepting a criminal record or submitting to "treatment." The latter alternative, tied as it is to criminal justice, has at times bastardized and denigrated the therapeutic aspect of the mental health system and has also had a profound effect on how the individual user functions in society and how he views himself.

After the Shafer Commission's final report,[51] about a dozen states partially decriminalized marihuana, that is, they reduced the first-offense penalties for possession of small amounts for private use to a fine similar to that for an expensive parking ticket, carrying no criminal record, and made the selling of marihuana the chief offense. Yet few authorities, including the Shafer Commission, believed that decriminalization was a viable, long-term resolution of the marihuana problem. It was intended to be experimental. It would buy time while society attempted to deal with the health effects and social consequences of marihuana use.

Unfortunately, in spite of the Shafer Commission's wisdom in some areas, it reflected what has become a traditional response to drug use. It sought to delay policy decisions in favor of further research, as has the more recent *Report of a Study by a Committee of the Institute of Medicine,*[52] implying that a search for more facts would produce clear and complete answers to policy questions—that new data would in some magical way eliminate the need for difficult intellectual or moral choices.

Yet the experiment of decriminalization did provide interesting data. Studies were done in several states that decriminalized, most notably Oregon,[53] California,[54] and Maine.[55,56] They indicated that in those states the use of marihuana had not increased at a significantly greater rate than in states that had not decriminalized, and moreover that law enforcement resources had been freed to attend to more serious criminal activities. But these findings did not arouse any great interest in policy changes that would lead to the establishment of formal social controls short of total prohibition.

Instead, the reverse occurred. Several new research studies appeared contending that marihuana presented greater health hazards than had been expected. The validity of these projects—except, perhaps, for those showing that this drug may cause lung damage as severe as that resulting from the use of tobacco and that it is probably bad for heart patients—is highly debatable and has been challenged by the authoritative *Report of a Study by a Committee of the Institute of Medicine.*[57] However, a finding more critical in inhibiting the move away from total prohibition was the discovery that a serious drop had occurred in the age of first use of marihuana, as well as an increase in heavy use among the lowest age group. The outcry about these findings led to the formation of antimarihuana groups who sought to "save" their children by antimarihuana campaigns for "education and prevention of any use," advocating stricter penalties and more law enforcement. These groups effectively exerted pressure on political officials.

The current marihuana issue is important to this discussion on controlled use of illicit drugs because it confirms one of our research findings—that without formal social controls to buttress informal controls, the age of first use could be expected to drop. It is interesting, however, that in 1979 and continuing in 1980 and 1981, the age of first use of marihuana began to rise again; whether from the increased activity of parents' groups or from the more effective functioning of informal social controls is a matter for later evaluation.

There is no doubt that the relationship between formal and informal controls is astonishingly complex. The two basic types of formal control are those provided by law and those enforced by controlling institutions, and these differ in specific respects. A high school, for instance, in a certain state, can forbid and punish the consumption of alcohol by all students, including eighteen-year-olds, at a senior prom even if that state approves drinking at age eighteen, just as a high school in a state that has decriminalized marihuana can still expel a marihuana user.

Informal controls are buttressed by formal controls. For example, in a state where drinking is legal at eighteen, an eighteen-year-old may give a beer to a sibling who is slightly younger, say seventeen; but he would rarely give a drink to a thirteen-year-old sibling. The interaction of these different forms of controls is most crucial in the case of young adolescents because everyone agrees that this group should not use any intoxicants at all, either licit or illicit.

At present, it is difficult to set standards (formal or parental) for illicit drugs because all such use takes place underground. Many parents have said that they can approach their children more easily about cigaret smoking than about smoking marihuana because, as one parent put it, "We can at least talk about cigarets—I can bribe, wheedle, cajole, or threaten. But with illicit drugs there is a code of silence. I'm afraid that attitude may spread to alcohol, which we used to be able to talk about."[58] Parents today are in a very difficult position. To the extent that the aim of informal social controls is safety and the prevention of problems, this matches the parents' aim. In many homes where alcohol is consumed, the formal controls (state laws) support the parents' sanctions and rituals that will aid in controlling use.

In the case of illicit substances, the situation is much less straightforward but institutional controls can still help. A secondary school that insists on and enforces formal controls, such as banning the illegal use of both alcohol and drugs while at the same time attempting a reasonable, low-key educational program on intoxicants, strengthens the hands of parents, who cannot then be told by youngsters anxious to experiment with marihuana that "it must be OK at any time and any place because even the school doesn't make a fuss." Such attempts at regulation at least give the parents the opportunity to think through with their children some of the informal controls that will help them to establish boundaries for safe use.

Policy Recommendations

Although I am a critic of the present prohibition policies, I differ from many other critics in not proposing that all illicit drug use be decriminalized or legalized. A typical example of that stand is Thomas Szasz's *laissez-faire* approach.[59] However, as John Kaplan and Mark H. Moore have pointed out in their recent articles,[60,61] the increase in the number of drug users that would result from such an approach would mean an increase in at least the absolute number of drug casualties. Because of this risk, a more cautious approach to policy change is needed than is provided by Szasz and most other critics.

First, the use of illicit drugs should be generally discouraged.

Second, all possible efforts should be made—legally, medically, and socially—to distinguish between the two major types of drug use: moderate, occasional use, which incurs only minimal social costs, and the more dysfunctional and compulsive use patterns.

Third, greater attention should be paid to conditions of use than to the prevention of use: to the conditions in which dysfunctional use occurs and to the conditions that promote controlled use.

Fourth, the development and dissemination of informal social controls (sanctions and rituals) among those who are already using drugs should be encouraged.

Informal social controls cannot be provided ready-made to users by well-meaning professionals, nor can they be created by formal policy, although there is a strong interaction between the two. They arise by largely unknown processes in the course of social interaction among drug takers under certain social conditions. They are not the work of the moment; they develop slowly in ways that fit changing cultural and subcultural needs.

This is the primary reason why any abrupt shift in present policy would be inappropriate. The sudden legalization of even marihuana might leave some users ill-equipped to handle its use because they might not yet have developed the social controls needed to regulate use and prevent abuse. There are, however, several cautious steps that could be taken now to demystify drug use and thus to encourage the development of appropriate rituals and social sanctions.

Awareness of the existence of these sanctions and rituals—the informal social controls—and their interaction with formal social controls is the first step. Because of the spontaneous and unconscious development of informal controls, this awareness cannot lead to rigid codification or to clear prescriptions. It can lead to educational policies aimed more specifically at dysfunctional use and carried out for the broad purpose of social learning rather than the narrow attempt to disseminate specific material in a classroom. Thus, other necessary ingredients in our effort to avoid the serious consequences of intoxicant use, such as treatment programs, legal reform, and medical research, should focus on the same problem and would lead to further cautious steps to reform our current policies and practices.

References

[1] N. E. Zinberg et al., "Patterns of Heroin Use," *Annals of the New York Academy of Sciences* **311** (1978), 10–24.

[2] T. S. Szasz, *Ceremonial Chemistry: The Ritual Perception of Drugs, Addicts and Pushers*, Garden City, New York: Anchor Press/Doubleday, 1975.

[3] M. Heller, *The Sources of Drug Abuse*, New York: Addiction Services Agency, 1972.

[4] N. E. Zinberg, *Drug, Set, and Setting: The Basis for Controlled Intoxicant Use.* New Haven, Yale University Press, 1984.

[5] *Marijuana and Health. Report of a Study by a Committee of the Institute of Medicine, Division of Health Sciences Policy*, Washington, DC, National Academy Press, 1982.

[6] E. Josephson, "Trends in Adolescent Marihuana Use," in *Drug Use: Epidemiological and Sociological Approaches*, E. Josephson and E. E. Carroll, eds., Washington, DC, Hemisphere Publishing Corporation, 1974.

[7] *Marihuana and Health. Sixth Annual Report to the Congress from the Secretary of Health, Education, and Welfare, 1976.* DHEW Publication No. (ADM)77–443, Washington, DC, US Government Printing Office, 1977.

[8]L. N. Robins et al., "Vietnam Veterans Three Years after Vietnam: How our Study Changed our View of Heroin," in *Problems of Drug Dependence*, L. Harris, ed., Richmond, Virginia: Committee on Problems of Drug Dependence, 1977.

[9]C. P. Herman and L. T. Kozlowski, "Indulgence, Excess, and Restraint: Perspectives on Consummatory Behavior in Everyday Life," in *Control over Intoxicant Use: Pharmacological, Psychological, and Social Considerations*, N. E. Zinberg and W. M. Harding, eds., New York, Human Sciences Press, 1982, 77–88.

[10]*Drug Use in America: Problem in Perspective*. Second Report of the National Commission on Marihuana and Drug Abuse, Washington, DC, US Government Printing Office, 1973.

[11]N. E. Zinberg and J. A. Robertson, *Drugs and the Public*, New York, Simon and Schuster, 1972.

[12]N. E. Zinberg, W. N. Harding, and M. Winkeller, "A Study of Social Regulatory Mechanisms in Controlled Illicit Drug Users," *Journal of Drug Issues* **7**, 1977, 117–133.

[13]N. E. Zinberg, R. C. Jacobson, and W. M. Harding, "Social Sanctions and Rituals as a Basis for Drug Abuse Prevention," *American Journal of Drug and Alcohol Abuse* **2**, 1975, 165–181.

[14]N. E. Zinberg and J. V. DeLong, "Research and the Drug Issue," *Contemporary Drug Problems* **3**, 1974, 71–100.

[15]D. Maloff et al., "Informal Controls and Their Influence on Substance Use," in *Control over Intoxicant Use: Pharmacological, Psychological, and Social Considerations*, N. E. Zinberg and W. M. Harding, eds., New York, Human Sciences Press, 1982.

[16]Zinberg, Harding, and Winkeller, op. cit.

[17]W. M. Harding and N. E. Zinberg, "The Effectiveness of the Subculture in Developing Rituals and Social Sanctions for Controlled Drug Use," in *Drugs, Rituals and Altered States of Consciousness*, B. M. du Toit, ed., Rotterdam, Netherlands, A. A. Balkema, 1977, 111–133.

[18]Zinberg, op. cit.

[19]H. N. Boris, N. E. Zinberg, and M. Boris, "Social Education for Adolescents," *Psychiatric Opinion* **15**, 1978, 32–37.

[20]*Drug Abuse Films*, Washington, DC, National Coordinating Council on Drug Education, 1973.

[21]US Cabinet Committee on Drug Abuse Prevention, Treatment, and Rehabilitation. *Recommendations for Future Federal Activities in Drug Abuse Prevention*. Report of the Subcommittee on Prevention, Washington, DC, US Government Printing Office, 1977.

[22]P. M. Wald and A. Abrams, "Drug Education," in *Dealing with Drug Abuse: A Report to the Ford Foundation*, New York, Praeger Publishers, 1972.

[23]M. S. Goodstadt, *Myths and Mythologies in Drug Education: A Critical Review of the Research Evidence. Research on Methods and Programs of Drug Education*, Toronto, Addiction Research Foundation of Ontario, 1974.

[24]B. Swift, N. Dorn, and A. Thompson, *Evaluation of Drug Education: Findings of a National Research Study of Effects on Secondary School Students of Five*

Types of Lessons Given by Teachers, London, Institute for the Study of Drug Dependence, 1974.

²⁵Yankelovich, Skelly, and White, Inc. *Students and Drugs: A Report of the Drug Abuse Council*, Washington, DC, Drug Abuse Council, 1975.

²⁶R. E. Mogar and C. Savage, "Personality Change Associated with Psychedelic (LSD) Therapy: A Preliminary Report." *Psychotherapy Theory and Research Practice* **1**, 1954, 154–162.

²⁷E. S. Robbins, W. A. Frosch, and M. Stern, "Further Observations on Untoward Reactions to LSD," *American Journal of Psychiatry* **124**, 1967, 393–401.

²⁸A. Huxley, *The Doors of Perception*, New York, Harper and Row, 1954.

²⁹A. T. Weil, *The Natural Mind*, Boston, Houghton Mifflin, 1972.

³⁰Robbins, Frosch, and Stern, op. cit.

³¹*Drug Use in America*, op. cit.

³²W. H. McGlothlin and D. O. Arnold, "LSD Revisited: A Ten-Year Followup of Medical LSD," *Archives of General Psychiatry* **24**, 1971, 35–49.

³³H. L. Barr et al., *LSD: Personality and Experience*, New York, Wiley-Interscience, 1972.

³⁴H. S. Becker, "History, Culture, and Subjective Experience: An Exploration of the Social Bases of Drug-Induced Experiences," *Journal of Health and Social Behavior* **8**, 1967, 163–176.

³⁵Robins et al., op. cit.

³⁶N. E. Zinberg, "Heroin Use in Vietnam and the United States: A Contrast and Critique," *Archives of General Psychiatry* **26**, 1972, 486–498.

³⁷Robins et al., op. cit.

³⁸C. P. O'Brien, "Experimental Analysis of Conditioning Factors in Human Narcotic Addiction," *Pharmacological Review* **27**, 1975, 535–543.

³⁹C. P. O'Brien, personal communication, 1978.

⁴⁰N. E. Zinberg, "The Search for Rational Approaches to Heroin Use," in *Addiction: A Comprehensive Treatise*, P. G. Bourne, ed., New York, Academic Press, 1974.

⁴¹L. N. Robins and J. E. Helzer, "Drug Use among Vietnam Veterans: Three Years Later," *Medical World News*, October 27, 1975.

⁴²Robins et al., op. cit., 1977.

⁴³N. E. Zinberg, "G.I.'s and O.J.'s in Vietnam," *The New York Times Magazine Section*, December 5, 1971.

⁴⁴*Accident Facts—1981*, Chicago, Illinois, National Safety Council, 1982.

⁴⁵N. E. Zinberg, "Alcohol Addiction: Toward a More Comprehensive Definition," in *Dynamic Approaches to the Understanding and Treatment of Alcoholism*, M. H. Bean and N. E. Zinberg, ed., New York, Free Press, 1981, 97–127.

⁴⁶*Marihuana: A Signal of Misunderstanding*. First Report of the National Commission on Marihuana and Drug Abuse, Washington, DC, US Government Printing Office, 1972.

⁴⁷*Drug Use in America*, op. cit.

⁴⁸Domestic Council Drug Abuse Task Force, *White Paper on Drug Abuse: A Report to the President, September 1975,* Washington, DC, US Government Printing Office, 1975.

⁴⁹Strategy Council on Drug Abuse, *Federal Strategy for Drug Abuse and Drug Traffic Prevention,* Washington, DC, US Government Printing Office, 1976.

⁵⁰*Report to the President,* from the President's Commission on Mental Health, Washington, DC, US Government Printing Office, 1978.

⁵¹*Drug Use in America,* op. cit.

⁵²*Marijuana and Health,* op. cit.

⁵³*Marihuana Survey—State of Oregon,* Washington, DC, Drug Abuse Council, January 28, 1977.

⁵⁴*Impact Study of S.B. 95,* California Health and Welfare Agency, Office of Narcotics and Drug Abuse, 1976.

⁵⁵*An Evaluation of the Decriminalization of Marihuana in Maine,* Augusta, Maine, Office of Alcoholism and Drug Abuse Prevention, 1978.

⁵⁶*Maine: A Time/Cost Analysis of the Decriminalization of Marihuana in Maine,* Augusta, Maine, Office of Alcoholism and Drug Abuse Prevention, 1979.

⁵⁷*Marijuana and Health,* op. cit.

⁵⁸Zinberg, op. cit., 1984.

⁵⁹Szasz, op. cit.

⁶⁰J. Kaplan, "Practical Problems of Heroin Maintenance," in *Control over Intoxicant Use: Pharmacological, Psychological, and Social Considerations,* N. E. Zinberg and W. M. Harding, eds., New York, Human Sciences Press, 1982, 159–172.

⁶¹M. H. Moore, "Limiting Supplies of Drugs to Illicit Markets," in *Control over Intoxicant Use: Pharmacological, Psychological, and Social Considerations,* N. E. Zinberg and W. M. Harding, eds., New York, Human Sciences Press, 1982, 183–200.

Controlling the Uncontrollable

John P. Conrad

Introduction

The obvious point of departure for any consideration of narcotics control is a truism that is honored entirely in the breach. Every official of the criminal justice system, every informed observer, and every citizen endowed with rudimentary common sense knows that the use of the criminal law to enforce standards of behavior is futile if substantial numbers of citizens do not wish to comply with those standards. To create a large public that accepts as legitimate the regular violation of the law is ruinous of law observance and corruptive of social and political institutions. Yet in spite of this common knowledge of the dangers inherent in governmental attempts to do the impossible, this nation, as well as nearly every other nation, both East and West, persists in the use of the police powers of the state for the purpose of restraining the traffic in narcotics. After a generation of effort, that traffic still eludes control. It is time to consider seriously the range of alternative controls that might limit the most undesirable effects of narcotics use with less destructive impact on the maintenance of order.

The principle just enunciated is no untested hypothesis. In 1931, the National Commission on Law Observance and Enforcement (more commonly known as the Wickersham Commission, after its chairman, George Wickersham, in his day a familiar eminence, though now faded into the obscurity of the archives), put it this way after a review of the nation's experience with the Eighteenth Amendment:

49

> It is axiomatic that under any system of free government a
> law will be observed and may be enforced only where and to the
> extent that it reflects or is an expression of the normally law-
> abiding elements of the community . . . The state of public opin-
> ion, certainly in many important portions of the country, pres-
> ents a serious obstacle to the observance and enforcement of the
> national prohibition laws.[1]

In spite of the lessons we should have learned as a nation from our experiment with the prohibition of alcoholic beverages—in itself a violation of the Commission's axiom—we are reliving that experience in an effort to prohibit the sale and use of narcotics. The consequences are even more disruptive of order. To remedy this state of affairs will call for conceptual vision, political courage, and administrative skill of a very high order. It is time to mobilize these resources. In this essay, I hope to lay some of the groundwork.

I shall begin by recalling the national miseries brought on by the Eighteenth Amendment, outlining the consequences of that experiment for the administration of justice. I shall then review briefly some of the harm done by persisting in our present policies. Drawing on the Hastings Center studies of "performance-enhancing" and "pleasure-enhancing" drugs, I shall propose an alternate strategy of control. I shall conclude that although the regulation of a licit traffic in narcotics will be subject to abuses, it will produce benefits that the absolute prohibition of that traffic cannot bring about, and it will also remove some of the most destructive consequences of the present laws.

I must preface these remarks with the recognition that, in the light of present-day public opinion, what I have to propose will be extraordinarily difficult to put into effect. Even if some bold and persuasive leader can turn the American people from their foolish course, the results will not put us on a road to Utopia. Many people will persist in using drugs who should not. There will be abuses of regulations and there will be outright violations that will have to be punished under the criminal law. Candor and realism require these concessions, but I will argue that my proposals will lead to a more satisfactory solution than the intolerable situation that now confronts the nation. One of the worst features of this situation is the lack of serious discourse concerning fundamental strategies of control. It is as though it were somehow immoral to consider seriously any alternative to an implacable policy of prohibition. What I have to say here is a contribution toward opening a virtually closed discussion.

The Lessons of a Failed Experiment

The sorry tale of the Eighteenth Amendment and the Volstead Act is still within living memory. Many accounts of its administration have been written but a comprehensive and objective history has yet to be compiled. The Amendment itself was ratified in January, 1919, and was in effect until the ratification of the Twenty-First Amendment, which repealed it, in December 1933. During the intervening fifteen years, the cities of America saw the growth of an illegal traffic in liquor, wine, and beer to dimensions which could only be estimated in the loosest fashion. Late in the day, Malvern H. Tillitt made a credible attempt to calculate the annual economic costs.[2] He thought the costs, as they stood in 1931, were as follows:

Enforcement of the law:	$50,000,000
Loss of revenue:	
(Taxes not collected at all levels of government from legitimate procedures)	$1,000,000,000
Excess profit taxes not collected from bootleggers:	$1,000,000,000

In addition, Tillitt called attention to the "hundreds of millions" lost to the economy by the abolition of employment in the production and distribution of alcoholic beverages. All these calculations were in the dollars of a time when the support of the federal government required an annual budget of less than $10,000,000,000. Prohibition was a formidable national effort. As the burden grew, the willingness of the people to support it understandably declined.

As dislocating as the economic damage was, the damage to our political institutions was much more serious. The Volstead Act created an uncontrolled market of vast proportions. Rival distributors of bootleg products depended on violence and threats of violence to maintain, and, if possible, to increase their share of the profits. They found it necessary to divert a portion of the gross profits to the corruption of the police and other public officials. One documented example will suggest the dimensions of the problem. It is drawn from the early days of prohibition, but conditions worsened over the decade and a half of increasing futility.

George Remus was a properous Ohio lawyer practicing criminal law with an income of $45,000 a year when, in 1920, he decided to abandon the bar. He purchased, for $10,000, a distillery in Cincinnati that was licensed to produce medicinal whisky under an exception allowed by the Volstead Act. His gross profit was $2,000,000 in the first year of business; it rose to $25,000,000 by the end of the third year. Years later he recalled for interviewers that the heaviest item of overhead expense was the "pay-off to local enforcement agents, storekeepers . . . and federal officials. (My) bribery costs were $19–$21 per case, but a case sold for $75 to $90."

Quite early in this new career, George Remus' prosperity attracted the attention of the Internal Revenue Service. The agent in charge of the investigation reported that "Remus had 44 people in his office (on one day alone) and some of them were federal prohibition agents or federal marshals. He paid them an average of $1,000 apiece." This information went to Washington, but no action was taken with respect to either party to these transactions. Remus' connections with the "Ohio gang," then so influential in Washington protected him throughout the early twenties. He later observed, "I tried to corner the graft market to find out that there is not enough money in the world to buy up all the public officials who demand a share of the graft." Eventually the Ohio gang itself was dispersed by criminal prosecutions and Remus was arrested, convicted of violations of the Volstead Act, and sentenced to two years in prison and a fine of $10,000.[3]

That was early in the day. Regular understandings had yet to be achieved among the various parties involved in the liquor traffic and its facilitation. As the years went by, monopolies consolidated the rival gangs of entrepreneurs. Increasingly the large urban states left enforcement of the liquor laws to the federal government. The result was described by John D. Rockefeller, Jr., originally an advocate of Prohibition, in a message to the Republican National Convention of 1932, using language that was close to apocalyptic:

> . . . drinking has generally increased . . . the speak
> easy has replaced the saloon . . . a vast army of law-

breakers has been recruited and financed on a colossal scale. . . Many of our best citizens . . . have openly and unabashedly disregarded the Eighteenth Amendment; that as an inevitable result respect for all law has been greatly lessened; that crime has increased to an unprecedented degree—I have slowly and reluctantly come to believe. . .⁴

Despite Rockefeller's jeremiad, President Hoover was renominated by the Republican party and campaigned on a platform that promised among other things, to continue the "noble experiment," as he was pleased to call the Eighteenth Amendment. He was decisively defeated, although it must be allowed that other factors in the political and economic situation of the time contributed to that result. The repeal of Prohibition—after that election a virtually dead letter—was speedily ratified during the first year of the Roosevelt administration.

The failure of Prohibition hardly needs an explanation, but its lessons are so relevant to the enforcement of laws prohibiting traffic in illegal drugs that some discussion is in order. The criminal justice system exists to enforce laws that interdict crimes against the state, the person, private property, and public order. It functions best in the enforcement of laws against the traditional crimes that have existed throughout the history of organized societies. Even criminals agree on the criminality of murder, rape, and theft. Agreement will never be so general with respect to behavior more recently defined as criminal. When the law is augmented with new crimes, disobedience will be inevitable on the part of persons who see nothing criminal in the behavior prohibited.

Disobedience is compounded when the crime is victimless. Where there is consent to the criminal transaction by all who are party to it, but no victim to complain, the police face difficulties that usually must be resolved by the use of paid informers and other surreptitious means. In effect, the police often must solve these crimes, if they are to be solved at all, by the corruption of a criminal. During Prohibition the tables were often turned; the police in many cities were corrupted, beginning at the top. Subterranean alliances were made between organized crime and politicians; some of these alliances survived the repeal of the Eighteenth Amendment for many years.

Otherwise law-abiding citizens patronized speakeasies and bootleggers during the twenties without compromising their inhibitions against the commission of the more recognizable crimes that have been embedded in the criminal codes for millennia. Their patronage of the illegal system for marketing liquor contributed to the formation of an uncontrolled and uncontrollable apparatus for merchandising a product that was in great demand, but which was obtainable only from committed criminals. In doing so, ordinary citizens made vast profits available to persons willing to accept the risks and stigma of a criminal career. That was obvious to any casual observer of the society of that era, and advocates of Prohibition seldom failed to make that point in their demands for more stringent enforcement by the authorities and more willing compliance with the law by citizens.

What was not so obvious was the organizational experience gained by shrewd criminals in the operation of a network of illegal monopolies. The powerful structure of organized crime that resulted has flourished ever since at a level of prosperity unknown in any other country. Other factors contributed to the organization of crime, but it is not conceivable that such a confederation of criminals could have attained its present influence without the 15 years of managerial experience and monetary profits gained under the auspices of the Eighteenth Amendment.

By no means all the ills of the criminal justice system in the eighties can be traced to the damage done in the twenties, but it is certain that the damage inflicted by our national imprudence was grievous and it is still unrepaired. For the last 30 years, we have been compounding the damage by our attempt to prohibit the manufacture, sale, and use of various psychoactive drugs. The theme is the same. Just as the Puritan strain in our culture demanded a perfect state of abstention from alcohol in the early years of this century, a similar demand is made for abstention from narcotics. Both goals are commendable and desirable; it is impossible to argue for a utilitarian justification of the use of either alcohol or narcotic drugs. Achievement of these goals, if they are to be achieved at all, requires a strategy of regulation and education. Individuals cannot be coerced into goodness, and neither can an entire nation.

New Lessons to Learn

Enforcement of the Volstead Act was simple compared to the difficulties encountered in our national drive to prohibit the marketing of illegal drugs. In spite of enormous expenditures of public funds (approaching $1,000,000,000 annually for the federal government alone), the diversion of police attention to the enforcement of these laws, and the investment of some remarkable tactical ingenuity and technological innovation, it has been a losing battle with heroin and cocaine and tacit surrender to marijuana. The results are discouraging, but the nation perseveres in the struggle to enforce these virtually unenforceable laws.

The same noxious effects that were so evident during the Prohibition era are once again ubiquitous. Enormous sums of money are accumulated from the inelastic market for heroin and cocaine. Much of it goes to mysterious figures in other countries beyond the reach of our police authorities. The market for marihuana flourishes to the extent that in some localities it is the highest earning cash crop. Vast funds have thus become available for the creation of criminal empires whose resources make them uncomfortably adroit antagonists for the forces of law and order. Occasional victories against these antisocial conglomerates do take place, but it is usually estimated that the drugs that are confiscated amount to no more than a tenth of those that reach addicts and users.

The dimensions of the problem can be glimpsed, though not accurately assessed in the figures in Table 1.

Methods for calculating the numbers of addicts and users are necessarily inexact. Using data compiled from reports of

Table 1
Street Value of Contraband Narcotics
Seized by the US Coast Guard[a]
1973–1979

Year	Value in millions of $
1973	4.79
1974	37.39
1975	34.80
1976	146.42
1977	429.59
1978	1319.58
1979	1990.42

[a]*Source: Source Book of Criminal Statistics, 1980.* Adapted from Table 4.28, p. 380, (Washington, DC, US Department of Justice, Bureau of Justice Statistics, 1981).

heroin treatment clinics and mixing in an element of informed speculation, Hunt and Chambers arrived at an estimate of the national population of heroin users alone that rose from 1,167,000 in 1968 to 4,350,000 in 1974. In supporting their figures, the authors explained that,

> Support for (these) figures comes from crude indications of the enormous size of the U.S. heroin market, as measured by seizures. In fiscal year 1974, state and federal agents confiscated about 5,500 pounds of heroin equivalents, most of which was probably intended for U.S. users. If this seized quantity amounted to ten percent of the total market, it would imply about 4.5 million active users at an average consumption of 6,000 milligrams per year (pure heroin). These estimates are, of course, at least as arguable as the figures we are trying to corroborate . . . They are, however, not unreasonable, and they imply an order of magnitude of users similar to the other estimates. The question is not whether there are three or four million, but that the number is several million rather than only several hundred thousand.[5]

The estimate made by these writers that the quantity of heroin seized is only ten percent of the total market may understate the true situation. That is an unknown which can only be guessed, but no one contends that seizures come close to a third of that total. The war on heroin is no better than an unstable containment.

By insisting on the prohibition of narcotics sales and use, we have denied ourselves the benefits of a regulated market. The development of regulatory structures for effective control has become immeasurably more difficult because of our perseverance with the prohibition of all narcotics sales and use. Formal control of the use of narcotics through the sanctions of the criminal law must come to an end. It should be replaced by a system of regulations supplemented by intensive public education. This transition will be difficult. The difficulties will not diminish by prolonging the delay. More years of futility and defeat of government efforts will entrench organized crime still further. The passage of time will also serve to increase the public's conviction that nothing can be done about this obstinate problem except to perservere in the present hopeless course. In that climate of opinion a destructive cynicism will flourish. The obstacles to constructive change will grow with the nation's perseverance in futility.

The Performance-Enhancing Drugs

To gain first-hand knowledge of the use of drugs to enhance performance in sports, the Hastings Center engaged the consultative services of professional athletes and sports physicians. A brief recapitulation of their account of the use of drugs is in order at this point.

Our consultants had varying opinions about the value of drugs in the different sports in which they had been engaged. The professional football players were certain that there are no drugs that can improve the performance of a player; indeed, one consultant—a well-known wide receiver—told us that psychoactive drugs would interfere with the precision needed for the execution of plays. For professional football, the usefulness of drugs is in the relief of pain. For that purpose, drugs are essential to performing at all.

Here a clear-cut ethical problem arises, although, strictly speaking, the question is whether performance-*enabling* drugs should be permitted. Team physicians have become so proficient in the temporary relief of pain that football players may continue to play even though they have suffered injuries so severe that patients in ordinary occupations would be disabled for weeks. When team physicians administer treatment in such a situation, they act in the interest of the club and in the short-range interest of the player. It is clear that the conventional physician–patient relationship has been compromised. Malpractice litigation may eventually establish boundaries for this questionable branch of medical practice. What may be more difficult to control is the self-medication by players anxious to continue their chosen profession, even at great risk of permanent disability or death.

This is not the problem that confronts us in considering the performance-enhancing drugs. Even the prescription of a pain-killer to enable an injured player to return to the field of play can be justified on the principle of informed consent. The use of pain-killing drugs by football players is an expected feature of their profession, just as the hazards of a permanently disabling injury must be accepted. The football player who understands the risks and accepts them is within his rights if he proceeds against all the counsel of common sense.

In individual athletic contests where precise measurement of performance can be made in time or distance, there is good reason to believe that some drugs improve performance.

Sprinters sprint faster and weight-lifters lift heavier weights. Undesirable side-effects are to be expected, and they are rejected by some of the consultants who discussed their experiences with us. These side-effects are part of the cost of excellence, but they are universally discouraged in official regulations. Where sports are conducted on a nominally amateur basis, the authorities will make strenuous efforts to prevent the use of drugs before contests take place. These efforts are so vigorous that sometimes athletes under prescribed medication—as, for example, asthmatics—are disqualified from competition.

A similar problem is the use of cocaine by basketball players in the professional leagues. It is thought that cocaine staves off fatigue and heightens alertness, though virtually nothing is known for sure about the validity of this notion.

So far as I can make out from information gleaned from a world to which I am a stranger, there is emphatic disapproval of the use of drugs in amateur sports and nominal disapproval in professional sports. Professional athletes do as they please so far as their clubs are concerned. We hear of athletes arrested for the possession of cocaine and marijuana, but we do not find them disqualified from play. [*See* Note Added in Proof.]

Sport is a unique occupation. Few other callings rely on drugs to relieve stress and to increase effectiveness. Considering the strain that most athletes must endure to continue competition during a season lasting many months, it is hard to dismiss the argument that performance enhancing drugs are necessities of the profession. I do not see how anyone can deprecate the use by football players of any drug that can be relied on to relieve pain. We heard that the informal code of professional football requires that players must play if they are not in a state of concussion and no bones are broken. Acceptance of pain-killing drugs is a part of the commitment that a player makes to his profession. We may deplore the long-range consequences of this physical abuse, but it is hard to see that the state has a definable interest in preventing a player from doing what he must to continue to play. Nor does it appear that the public has cause to complain if those who entertain the millions must anæsthetize themselves so that the show may go on.

If we are to accept the necessity of these drugs, there must be some regulation of their use. Players should understand the correct dosages for given types of injury, and they must be instructed on the consequences of overdosing.

As to the true performance-enhancers, I suggest that the policy should be *laissez-faire*. If we have departed from the amateur ideal of sport for sport's sake to the extent that national prestige depends on victories in athletic contests, we might as well pursue a policy of rational performance enhancement. A policy of informed consent—with official acceptance of the withdrawal of athletes who will not or cannot consent—is the course of realism. At present, athletes do not have reliable and authenticated information about dosages and consequences, and it may be supposed that sports physicians have much to learn from controlled research in these matters. Such investigations certainly cannot be undertaken until restrictions on the use of performance enhancers are greatly modified.

In summary, much will be gained by abandoning the formal prohibition of the use of drugs in sports. Athletes should be free to use drugs but they must know what they are doing to themselves. Sports physicians should have the benefit of serious research on the effects and hazards. This is an era when it is an implicit aspect of the athletic experience that drugs are needed to compete effectively in the popular spectator sports. Athletes who find this prospect repugnant—and surely there are many who do—can engage in the truly amateur sports that are good exercise, lots of fun, and require no drugs at all.

The Pleasure-Enhancing Drugs

I have known many people who have sought pleasure and surcease from care by using one or more of an array of pleasure-enhancing drugs—often at risk to their health, their personal liberty, and even to their lives. Most of this acquaintance found their way to prison; my sample is hardly representative. Those who use heroin, cocaine, and marijuana feel the need of these drugs so strongly that they are willing to accept these risks. Their objective is the physical experience of ecstasy—the famous heroin "rush," the transitory euphoria of cocaine, and the "stoned" condition of the cannabis smoker. Although drug users prefer company when they engage in these experiences, the experience itself is peculiarly isolating; it cannot be shared. The sensual experience is the aim. In seeking it, the drug-user removes himself from a reality of ennui, the humiliations of subordination and drudgery, and the lack of access to approved, conventional pleasures. It is little wonder that contemporary youth, at a loss to foresee satisfactions as adults in relationships or careers, so frequently search for oblivion to blot out the realities of alienation. The use of cocaine and marijuana are the answers that too many modern men and women have found for their discontents. To be "stoned" means more to the marijuana user than most of the rest of his experience. Until the prevailing alienation is relieved, we will have to accept the need that a substantial part of the population has for drugs to create the illusion of well-being. If this interpretation is correct, a lot of people believe they need drugs regularly and often, moralists and medical advisers to the contrary notwithstanding. It is hard to see why an adult should be denied this self-indulgence so long as he or she understands whatever hazards there are in the use of such drugs.

Instead of persistence in the pursuit of an impossible policy, our society should combine regulation of the product and its use with a campaign of realistic public education. Let us suppose that marijuana were legalized and the makings were available at reasonable prices in the open market. Minors would not be allowed to buy the stuff, just as they are now denied the right to buy liquor and tobacco. Of course, there would be a great deal of surreptitious use of such drugs by children. That happens under the present system, and it is impossible to prevent, even by strenuous efforts by the police to find and prosecute vendors. But if all the vendors of drugs are legitimate merchants, their sales practices can be regulated. Those who knowingly sell marijunana or other drugs to minors will face the loss of a valuable license or more serious sanctions.

It is unclear just how serious the hazards of marijuana are. Experiments on monkeys and rats are suggestive of some undesirable effects, but they are hardly conclusive. When marijuana use is legitimized, it will become possible to study the effects on human users with more rigor than now is the case. If research shows that the health of the user is substantially impaired, cannabis may lose favor. That is a goal that has eluded those who would prohibit its use. If it is relatively harmless, as some suspect, then the argument in favor of prohibition collapses, and we have only to consider whether any regulation at all is needed.

There are similar doubts about the effects of cocaine, heroin, and the rest of the pleasure-enhancing drugs. I do not doubt that much stricter measures are needed with respect to the control of the amphetamines, the barbiturates, and "angel dust"—or phencyclidine piperidine (PCP). Those pleasure-enhancers that have legitimate pharmacological uses should be strictly regulated if we have reason to believe that public health is endangered by their indiscriminate use. But where a drug can be used without serious danger to the user, as appears to be the case with cocaine and may also be true of heroin, the rational course is to combine regulation and public education, as we do with other drugs in general use. If adult citizens have had the opportunity to inform themselves of the consequences of drug use and choose to take whatever risks there may be, it is not possible to reconcile the state's interference with the principle that citizens should be allowed freedom to the extent that their actions do not infringe on the rights of others.

Reformulating the Ethics of Drug Use

I am impressed with the speculative quality of most of our beliefs about the effects of drugs on the body of the user. The majority of the public undeniably believes that the use of psychoactive drugs by anyone for any purpose other than the treatment of an illness or disability is an unmitigated evil, to be stamped out at all costs. Few of the millions who hold to this belief have used drugs or have had much contact with drug-users. They have for years been subjected to a barrage of prop-aganda consisting of palpable untruths mixed with uncertain and prejudiced conclusions reached on insufficient evidence. The ethical principle that results is simple: the use of drugs to enhance performance or pleasure is wrong and must be pre-vented. It is therefore the duty of the law to arrest, prosecute, and incarcerate those criminals who produce and market these drugs.

My argument will be apparent from the preceding pages. It is manifestly impossible to prohibit the use of these drugs or even to interfere effectively with their distribution. This view does not rest merely on the *a priori* review of the problem and the evident obstacles to its solution, nor on the historical sup-port provided by our experience with the Eighteenth Amend-ment. We have the benefit of a truly rigorous study of the im-

pact of the New York Drug Law of 1973, performed by the
Association of The Bar of the City of New York and the Drug
Abuse Council.[6] The law itself prescribed severe and manda-
tory penalties for all kinds of narcotic drug offenses. The study
drew on three years of experience with the statute and demonstrated that it failed to achieve its objective of reducing drug
use and interrupting traffic. Heroin use was as widespread in
1976 as it was in 1973. Serious property crime increased
throughout the state. The expense was considerable; about
$32,000,000 was spent during this period in the enforcement
and implementation of the law.

No more convincing demonstration of the futility of our
present laws can be found than in this evaluation of the impact
of increased severity. If what we are now doing does not accomplish what we want, we know that doing more of the same
will not accomplish more.

Adults should be free to decide for themselves whether or
not to use narcotic drugs. Their choice should be enlightened
by all the scientific knowledge about the effects of these drugs
that we can muster. If they can be safely used in properly cali-
brated doses they should be available. Those drugs that are
hazardous under any circumstances should be controlled like
any other poisons and kept off the market except for purposes
that can be approved. What we have now is an unregulated
market in which consumers use what they can get without
guidance except by hearsay and "street pharmacology" about
the precautions to be taken. What they can get is unpredictable
as to quality, availability, and cost; it may not even be the substance the consumer was promised. If heroin is temporarily out
of circulation, methadone will have to do. If cocaine cannot be
had because of scarcity or cost, "angel dust" may be sold as an
acceptable substitute.

A regulated market will provide cheaper drugs and drive
off the exotic and really dangerous substances. At present, society gets the worst of all worlds. The law is damaged by its
attempt to do what cannot be done. Those who use drugs are
cheated and exploited by dishonest dealers and pushers who
cannot be reached on charges of consumer fraud. Behind the
traffic is a powerful organization of criminal activities that gains
its ends by corrupting officials, defying the law, and
intimidating competition.

If the state should do no needless harm to its citizens, that
principle is violated by the futile struggle to maintain control

through the administration of the criminal law. Regulation of a product like narcotic drugs will eliminate the apparatus that organized criminals have created for the distribution of unstandardized drugs. The safety of the consumer can be protected, and the immense costs that have been incurred by subjecting this traffic to the criminal justice system can be wiped out.

A police chief of a large city with a gift for heavy irony once said to me that the nation should be grateful to the narcotics pushers who sedate the unemployed youth of the inner cities. Were it not for their tireless efforts, he said, the crime problem would be hopelessly unmanageable. Decriminalization of the narcotics business will relieve part of the crime problem, but it will not touch the roots of disenchantment. Until we can relieve the misery of redundancy, the narcotics problem will loom over us no matter what policy of control we may adopt. It may be the curse of the age as most people suppose, or it may be an unfamiliar blessing as some users ardently believe.

It will not disappear, it cannot be prevented, and the only hope for minimizing whatever harm it can do—or maximizing its benefits—must come from reasoned regulation. This transition will not be easy. Its completion will not be the end of all associated evils. Nevertheless, in a world in which the right is uncertain in so many respects, here is one aim that can be pursued with confidence that wickedness can be toppled and some good ends achieved.

References

[1]National Commission on Law Observance and Enforcement, *Report on the Enforcement of the Prohibition Laws of the United States,* Washington, DC, US Government Printing Office, 7 January 1931, pp. 44–60.

[2]Malvern Hall Tillitt, *The Price of Prohibition,* New York, Harcourt, Brace, 1932, p. 7.

[3]John Kobler, *Ardent Spirits: The Rise and Fall of Prohibition,* New York, Putnam, 1973, pp. 315–322.

[4]Id., pp. 350–351.

[5]Leon Gibson Hunt and Carl D. Chambers, *The Heroin Epidemics: A Study of Heroin Use in the United States, 1965–75,* New York, Spectrum, 1976, pp. 112–113.

[6]The Association of the Bar of the City of New York and the Drug Abuse Council, *The Nation's Toughest Drug Law: Evaluating the New York Experience,* New York, The Association of the Bar of The City of New York, 1977, pp. 9–12.

Note Added in Proof

Since this paragraph was written, drug scandals have erupted in professional football, baseball, and basketball, as well as in international track and field. Several famous players have been arrested and sentenced to jail terms. Others have been disqualified from play for brief periods of time. Measures have been taken by the sports establishment to discourage the use of narcotics and to "rehabilitate" some celebrated users. It remains to be seen how well these intentions will be translated into firm policy.

The State's Intervention in Individuals' Drug Use

A Normative Account

Robert Neville

The question of state intervention is a political one. What follows is an attempt to articulate two abstract political principles, to apply them to problems of state intervention in drug use, and finally to suggest an appropriate political point of view. The question of drug use is fraught with moral connotations, however: the mores of the societies within which drugs are used. Therefore I shall focus primarily on a question of judgment, namely, how "informed" a person should be in order to use drugs "responsibly." The relativities of judgment in determining what it means to be "informed" and "responsible" constitute important factors in establishing an appropriate political point of view.

First, it is necessary to circumscribe the meaning of "drug use." By "drug use" I mean the use of chemicals whose effects include noticeable psychological alterations toward which identifiable affective responses typically are made. That is, the psychological alteration is approved, desired to be repeated, enjoyed at the moment, appreciated as a relief from a previous state or as a change for the worse, feared, or construed as upsetting or damaging. The relevant affective responses may be those of the drug user, as in the case of a desired euphoric, or those of others, as in the case of the custodians of a violent psychotic calmed by a tranquilizer. It would be convenient to limit the discussion to "nonmedically sanctioned drug use" since that is where state intervention seems to be an issue. But medical prescription of drugs is itself a powerful form of state intervention. It is often recommended as a way of better intervening in cases of abused drugs that are currently nonmedical, e.g., heroin maintenance. Furthermore, in many respects the medical practices of dispensing minor tranquilizers and amphetamines bear closer analogy to bartending or the administration of sacred mushrooms than to medical practices of surgery or prescription of insulin or digoxin. So although the center of gravity in the following discussion is located among the drugs not usually associated with medical purposes, the domain of the discussion includes the use of any drug with noticeable psychological effects toward which identifiable affective responses typically are made.

Two Political Principles

The two political principles can be stated in tandem. Principle I, a principle of private autonomy deriving from the liberal tradition, is that *with proper qualifications there is no justification in the state's intervening in direct, responsible transactions regarding drugs between users and dispensers, or in the possession and use of drugs by individuals.* Principle II, an immediate qualification of the first, derives from John Dewey's political philosophy and is a principle of justified intervention. *The public consisting of those indirectly involved in transactions regarding drugs, their possession and use, by virtue of suffering the consequences of the transactions, possession and use, has a justified interest in intervening to control those consequences.*[1]

Before considering these principles in detail, some general observations are in order.

First, the principles seem on the surface to express the general liberal thesis that people can do what they want up to the point where they harm others, and then their activities are liable to regulation. I do believe this liberal sense is captured in the principles which, like liberalism, make political theory hang on the distinction between the public and private. But the principles extend beyond liberalism in that they focus on who can do what rather than on to whom something can be done, on the agent rather than the patient. The principles are united by the more general observation that people are justified in participating in any social process that affects them. Therefore, if individuals' drug behavior has consequences for other people, that is an invitation to the other people to intervene in the interests of controlling those consequences. The principles frame a theory of participation, with rights secondary to participation rather than the other typically liberal way around; the derivation of rights from participation is typical of pragmatism's political theory, as noted in footnote 1. Neither these principles nor liberal ones say directly what interventions should strive for or what drugs are good for; those values need to be defined contextually in terms of the individuals whose participation or intervention is justified.

Second, another general observation is that the proposed principles provide a contextual definition of who is justified in making interventions, namely those suffering the consequences of drug use. To move from such a narrowly defined public to the state requires an extra argument, which I shall make later.

Third, a final general observation is that there is a certain looseness of connection between the principles. That individuals' drug use has consequences for others does not entail that the others perceive the consequences, or that if they perceive them they care to do anything about them. It is perfectly justifiable for one society to intervene in a certain kind of drug use because people care to, whereas another society leaves the behavior in a nonpublic realm because no one cares to control its consequences. The principles, in other words, do not by themselves take a stand on how permissive a society should be.

Conditions of Responsibility

To spell out the significance of the principles for drug use in more detail, let us turn to the first, that with proper qualifications there is no justification in the state's intervening in direct, responsible transactions regarding drugs between users and dispensers, or in the possession and use of drugs by individuals. The controversial requirement in this is that the transactions, possession, and use be *responsible*. Let me explicate responsibility in this context by expanding on the argument given by the court in *Kaimowitz vs Dept. of Mental Health*.[2] The court there argued that a decision is responsible or free when three conditions obtain: (1) that the agent is relevantly informed, (2) that the decision is capably made by the agent without coercion, and (3) that the agent be competent or practiced at deciding. To this I would add that the agent be legally responsible so that the consequences of the drug use will not be the legal responsibility of someone else. Finally, I would observe that being responsible is itself a sufficiently important value that it would be irresponsible to use drugs in such a way as to diminish one's own responsibility in the long run, however justified it may be to enjoy a short-term holiday from responsibilities.[3] How do these dimensions of responsibility relate to drug use?

With reference to the first, the direct handling, possession, and use of drugs needs to be relevantly informed if it is to be responsible. This requirement unfortunately lends itself to stupid stipulations. The Kaimowitz court, for instance, seemed to argue that if a procedure were experimental, then one could not in principle know the relevant information, namely, the outcome of the experiment, in advance, and therefore could in principle never give informed consent. Let us suppose, however, with a caution deriving from historical perspective, that public discussion can give rise to a reasonable range of relevant information. It would seem that individuals could be expected to know the main possible consequences of drug use for themselves, scaled according to probability and importance. We are more aware than our grandparents, for instance, of the delayed effects of chemicals on health—as carcinogenic, for example, or as affecting fetuses when used by pregnant women—and responsible drug use should be informed about these possibilities. It seems too that individuals should be informed about the economic and social systems that make drugs available and that might be reinforced by their drug use. They should also be informed about the likely consequences of their drug use for others who might have an interest in intervening.

The problem with responsible information regarding drugs, however, is that there is sometimes a vital difference between knowing something and really believing it, between knowing a probability and accepting that it is probable for oneself. One's interest in using most psychoactive drugs is often an intensely personal one, whether it be to secure sensible pleasure, to avoid pain, or to groove on one's addiction; it is often unwise to trust what one believes about what one knows in cases of intense personal interest in the outcome. Yet to press this argument very far would fall into the trap of saying that anything personally important enough to be an issue of individual autonomy is likely to bias our capacities to accept knowledge so that it is generally impossible to be properly informed about any matters of intense personal freedom. To avoid this extreme conclusion, it is proper to say that individuals are responsibly informed if (1) there is the relevant information and (2) they have had access to it. The burden of proof should fall upon the argument that the individuals in particular lack the appreciation of the significance of the knowledge.

But should the case adequately be made that people are not responsibly informed, then the *prima facie* exemption from state intervention specified in Principle I would fail to hold. Suppose this is the case. What then? Liberalism would tend to hold that nothing happens then unless the principle of justified intervention in turn applies. But suppose that lack of responsible information does not lead to or alter the consequences of drug use on others. The principle of justified intervention then would not apply, and liberalism would say that that ends the matter. I would say that a further principle of social responsibility then might still apply, namely, that the social group as such has the responsibility to foster capacities for responsibility in its members. Perhaps the state has the responsibility to make informed judgment a condition for access to the use of drugs the way it does for access to a driver's license.[4] It is an empirical matter to determine whether increased responsibility regarding informed judgment can be developed. Then it becomes a question of social morality whether it is worth the costs.

Suppose, however, that in the case of the failure of informed judgment, over and above any cultivation of increased capacity for responsibility, drug use is harmful to the user (though not to others) in the opinion of those who are informed about these things. Does a principle of paternalism apply that would legitimate intervention to alter or stop the drug use? I believe the answer to this question is relative to different social groups according to the degree of organic tightness in their sense of social solidarity. In some cultures, being an esteemed and cared for member of the group is more important than freedom from the perhaps arbitrary, binding, and misguided dictates of the cultural authority. For those cultures based upon tight solidarity, principles of paternalism would indeed apply and would be thought of rather as principles of acceptance and membership. For cultures with the opposite balance, principles of paternalism would not apply, and I believe ours is a culture of this opposite sort. The diversity of subcultures and noncultures making up the American scene renders much cultural solidarity impossible, and the imposition of paternalistic values would have to be imperialistic except in accidental cases. So I would say that when a drug user's relevant judgment is uninformed, when it has no significant consequences for others, and leaving out of account the potentially educational value of intervention, the state still has no justification in intervening in order to prevent a person from harming himself or herself out of ignorance. Notice that this stand refers to cases of lack of informed judgment. Somewhat similar problems have yet to be raised with regard to capacity to choose without coercion and with regard to competence at making judgments, both of which may be impaired by drug use. Furthermore, an argument might be made that certain kinds of drug use without informed judgment by their very existence define the kind of society one has, and therefore constitute an indirect consequence others have an interest in controlling; certainly some voluntary organizations, such as churches, demand special kinds of private behavior from their members. I shall suggest at the end that the most a pluralistic society can demand is responsible behavior where the good of society is at stake.

Before leaving the topic of relevant information, however, we need to deal with the fact that regardless of the drug user's knowledge, drugs themselves differ in what is known about them by the best authorities. Marijuana, for instance, is not known now to be very harmful, whereas amphetamines are known to be harmful; and we can imagine that some newly invented Soma would raise very complicated questions about where to look in the individuals' and the society's life for potentially harmful and beneficial effects.

Regarding drugs whose effects are not known to be harmful, or where the harm is slight and easily understandable, no group would have a standing that would give them an interest in intervening. Some people operate according to a principle of caution, of course, under which nothing should be done until proved safe. But the burden of proof justifying the application of that cautionary principle on others seems to fall upon those who want to apply it; if individuals want to use drugs not known to very harmful, the risk would seem to be a matter private to themselves. The exception to this would be where there is a reason to believe that a possible, but as yet unknown, effect would have serious consequences for others, for instance, from a drug that may have the potential to cause a person to commit mayhem. Regarding drugs whose effects are known to be harmful, the situation is more complicated. The principle of justified intervention would dictate regulation of drug use whose harms fall directly on others, as in the case of mayhem, and an argument could be made justifying intervention when the harms fall indirectly on others, as in the case of others having to pay for illness caused by drugs, for instance in cigaret-caused cancer. But what about the case where the harm falls mainly on the user, with no significant consequences for others? If we lived in a purely liberal society where people could rot in privacy so long as they do not harm others, then no intervention would be justified. But it seems to me that among the things highly valued in our society, though perhaps less so than in traditional societies, is a habit of taking care of people so that they live within a range of locally accepted human amenities. Harmful drugs might be justified in being controlled, therefore, because their uncontrolled availability would allow users to sink to socially shameful depths of abject filth, mindlessness, and disorder. This rationale for the desire to control harmful drugs is coordinated with our attempts at drug rehabilitation programs for people whose lives fit slum environments. Part of the rationale for those programs of course is the interest to control the social consequences of criminality in the drug life; but that could be satisfied by making the desired drugs freely available. The British experience seems to have shown that although availability reduces crime, it does not seem to make much of a difference to the level of social amenities drug users sustain for themselves.[5] If we could eliminate the criminal costs to society, would we allow the abject victims of drug use to rot in their autonomy? I suspect not. We would hold the value of their enjoying the social amenities sufficiently important both to attempt to withhold the harmful drugs and to help them in their immediate situation. The reasons we cite for these attempts amount to saying that it is inhumane for society to let people victimize themselves to the extent common in the drug scene. This reflects not so much a principle of paternalism as a principle of social solidarity, of identification with the plight of the abject.

Now the sentiments I have just expressed would likely find strong support when applied to cases of users who were not convincingly informed about the known harmful effects of their drugs. But what about persons who are fully persuaded of the harmful effects and enter the drug life precisely because of them, out of some morbid desire for self-punishment or for a fascinating flirtation with death, or in order to punish others? Suppose these drug users are fully in accord with the principle of autonomy in that they responsibly, and with consequences only for themselves, choose a seriously harmful use of drugs. This situation may be a peculiar case of the right to commit suicide, which in some instances I would support. But, still, the value society holds for each individual being enclosed within the arms of society's amenities seems to me to justify a shifting of the burden of proof. Whereas ordinarily the burden of proof is upon the public to justify the intervention, here the burden of proof is on the individual to justify the use of the drug for self-destructive purposes. The individual's justification would have to take the form of justifying an exemption of himself or herself from solidarity with society regarding social amenities. Such a justification need not involve persuading the public that the exemption and self-destruction are good, but it would involve making a persuasive case that the individual can indeed abandon the society so that solidarity is no longer relevant.

Let us now recover the structure of the argument and recall that we have been adding detail to the requirement in the principle of autonomy that drug users be informed in their use. One more case needs to be raised in this regard. Suppose a genuine Soma drug is developed, one with no harmful effects and that is safe in dosage. Suppose also that equally benign drugs are developed to enhance perception, awareness, memory, and alacrity of thought. Only pharmacological Calvinists would object to these on intrinsic grounds.[6] But the special question of information is how the use of these drugs would alter individuals' lives and the texture of society, how much they would be used and by whom. By and large, the relevant kind of information of this sort is a matter for serious social study, not for individual self-knowledge. Would some become more than a desirable holiday drug, and become instead an occasion for nonproductivity beyond the tolerance of society? How would a society determine the proper tolerance of nonproductivity? With a good Soma, how much productive affluence would be needed? The answer to these kinds of questions involves considering many factors beyond the simple opposition of Soma versus productivity. But these answers are likely to be unobtainable without society's actual experience with the drugs in question. And even after such information is available in some form, there remains the knowledge of self that would be important to using these drugs. How much Soma one should take affects the overall contours of one's life the way how much liquor one should drink does; responsible ingestion requires considering this. Analogous points could be made about drugs altering ordinary capacities for perception, memory, and so on. But we must be wary of insisting on too much information and self-understanding, for otherwise we make it impossible to know enough to be free from liability to intervention.

Second, the other criteria of responsibility in drug use may be reviewed quickly since they present many of the same problems discovered in the requirement of informed judgment. The second requirement listed above is the capacity to use drugs without coercion. Two kinds of coercion seem relevant to mention: (1) the internal coercion of addiction and (2) the external coercion of pressure from others and from circumstances.

Addiction is notoriously hard to define, ranging from physical propensity to produce withdrawal symptoms to psychological inability to abandon the affective benefits of drugs.[7] In all these cases, however, addiction should have no legal force as a consideration of coercion if the means are readily available to eliminate the addiction. If people are addicted and want not to be addicted and have the means to kick the habit, then they are not coerced, I think. If they allow their addiction or more basic addictive character (if there is such a thing) to keep them away from the available means of relief, then that is a matter of their responsible choice, and not coercion.

External coercion from other people is more complex to understand. Some years ago we would have been inclined to say that the influence of others is not coercive unless they actually jab the needle in your arm. But recently, mainly under the influence of the women's movement, we have come to see that others exert extraordinary influence over us through the mediation of a system of social expectations, when we too have bought in to that system. Perhaps it is the case that the social expectations in a slum are such that younger children are highly defined by their expectations that they should try drugs when the older children tell them to. Nevertheless, many people can escape the system of expectations simply by rebelling against it when it is objectified. In fact not to rebel when the expectations are objectified is to take responsibility for holding them. Furthermore, the value our society places on personal responsibility justifies large interventions to make people aware of the systems of expectations about which they should be responsible; drug education programs are very good things, even when done badly, because they objectify the drug scene.

The cases where individuals fail to be responsible because they are truly coerced into drug use probably fall between forcible administration and cooptation by a system of expectations. If persons are rendered nonresponsible by virtue of coercion, and supposing that there are no serious consequences for others that would activate the principle of justified intervention, are there other principles that would justify an attempt to regulate drug handling, use, or possession? Certainly the criminal laws against harming others should be brought to bear on coercers. How could personal drug use or possession be coercive, however, except insofar as addiction or social expectations are concerned? With those exceptions, I doubt that any intervention could be justified.

Thirdly, the concept of drug addiction, unwieldy as it is, suggests that coercion for a short time leads by itself to coercion for a long time. This shifts discussion of the criteria of responsibility from capability of use without coercion to competence to make judgments regarding use, the third *Kaimowitz* condition. To be capable of exercizing responsibility requires a developed character for inhibiting impulse, figuring out consequences, organizing one's forces, and appreciating the moral self-definition that responsible choice involves. Capacity for drug use suffers the same untimeliness as the capacities for sex and violence: it arrives far earlier in life than capacities for competent relevant decision-making. The answer to the problem of drug abuse in children is clear, although well-nigh impossible. Other people, parents or surrogates, must exercise competent judgment in drug use for children and help them develop their own capacities for competent judgment. After controlling for such qualifications as special harms to the health of growing children, there is no reason to believe that drugs are less affectively meaningful for children than for adults. Of course in a society with very little disciplinary solidarity it is almost impossible to exercise parental restraint. But drug use presents few problems beyond those already resident in sexual behavior and violence.

Consider competence regarding drug use in adults. A democratic society runs a grave risk when it puts the burden of proof of competence on the individuals themselves. Besides the dangers of overscrupulosity analogous to those mentioned above concerning perfect information, having to prove one's competence in the face of a contrary presumption would involve too much snooping into private matters and too much personal record-keeping. Rather the burden of proof of incompetence should fall upon the public. It should be presumed that people are competent to decide about their lives when they reach, say, the end of high school, and that if they are not competent in some respect, the responsibility for developing that competence falls upon them rather than upon parents or surrogates. Proof of incompetence would have to consist either in citation of repeatedly incompetent judgment (incompetent, nor merely mistaken) or in some independent knowledge of impaired mental development or destruction. In the latter case, the state could intervene out of its interest in protecting the mentally deficient or ill (although we know how treacherous these waters are). In the former case it is not clear who would have standing to try to prove that someone is incompetent to choose about drug use and therefore should be controlled. If the incompetence leads to effects upon others that the others have an interest in controlling, the principle of justified intervention would define a relevant "public" appropriate to intervene. But if there are no serious consequences to others, the only justifications for intervening to save someone from incompetent judgment would be those stemming from solidarity that we noted in connection with saving people from being uninformed about drug use.

Fourth, a final requirement for responsible judgment in drug use is that the user be legally liable for procuring, possessing, and using. To the extent that parents or an institution would be legally liable instead, the freedom of the individual from intervention, even in cases where others are not seriously affected, cannot be defended. Of course this observation does not solve problems of prudence in making individuals morally responsible when legal responsibility resides in surrogates.

Conditions of Intervention

Let us now turn from consideration of drug use in light of the principle of autonomy to the implications of the principle of justified intervention, the principle, namely, that the "public" consisting of those indirectly involved in transactions regarding drugs, their possession, and their use, by virtue of suffering the consequences of the transactions, possession, and use, has a justified interest in intervening to control those consequences.

Definition of the Public

According to this principle the right to intervene falls upon those affected by the drug use and its accompaniments. Is a public defined that way sufficiently broad as to be coextensive with the groups whose agency of intervention is the government? I suspect that one of the corruptions of American democracy has been the deliverance to ever higher and more general levels of government the responsibilities for intervention that are much more local in their warrant. But in the case of drug use in America, I believe that the government is the intervening agency for the relevant public. This is for two reasons.

First, one of the major effects of drug use is the economic system or systems involved in making and distributing the drugs. Drug use is shaped by these systems, and it in turn reinforces them. There are elaborate international criminal systems of drug smuggling and distribution, international legal systems of drug manufacture and distribution, and criminal mismanagement and misappropriation in the otherwise legal drug industry. The scale of the economic base of drug use makes the entire citizenry the relevant public as represented in government.

Second, the direct effects of drug-oriented life styles are distributed according to proximity rather than any other more specialized kind of relation. The heroin addict and the tranquilized housewife affect those in their neighborhood, not merely those in their line of business, for instance. Government is the relevant agency for handling consequences of behavior that are distributed according to such anonymous or public forms of social interaction.

There are many justified interests in seizing control of the effects of drug use having to do with desirable economic systems and tolerable neighborhood life styles. I have only two observations to make in this regard. The first is that the personal and social costs of intervention are very high and therefore are among the most important values to be considered in weighing what to do. The second is that drug using often involves two roles, the personal role in which the individual's drug use might properly be protected by the principle of personal autonomy, and the social role as drug distributor within the economic system and as player of drug-oriented life styles within the society, which role might very well be liable to intervention and control.

Tasks for the Public

The great costs of intervention in the economic system of drug use are illustrated in the benefits organized crime enjoys from our laws protecting privacy. To do much better in stopping criminal drug traffic than French Connection-style heroic busts, the police would need surveillance techniques and license that no free society would tolerate. Or so it seems. In order better to regulate the legal drug industry and to educate physicians not to be bamboozled by greedy advertisements or pathetic pleas from patients wanting attention, efforts would be needed in government, media, and education that we seem unwilling to make. In light of this we must be very careful not to levy punishments against drug users out of frustration for our inability to intervene in the systems that supply them. The principle of justified intervention does not justify intervention against users as scapegoats, only in social processes that cause undesired consequences.

The dual role of drug users, however, makes the warning about scapegoats difficult to interpret. Many illegal drug users are also pushers, as they must be to support their own use. Of course the financial quantity of their selling is small compared with higher ups in organized crime, and the police would be

well advised to hit high for this reason. But then in some neigh-
borhoods, it is the little people, the user–pushers, who seduce
their friends and relatives into drug use. My own feeling is that
the person-to-person seductions into the drug life are not matters for the government, but for more local publics: the family
and individual group associations. But whatever the agent of
intervention, the dual role of pusher–user seriously compro-
mises the autonomy of possession and use by itself. Less dra-
matic but equally important is the dual role of legal drug misus-
ers. People who abuse minor tranquilizers, barbiturates, or
amphetamines for psychological purposes of enhancing lives
that otherwise are dull or ragged may hurt only themselves and
their nearest kin by what they directly ingest. But by supporting physicians as pill pushers, demanding medical time
and attention, and paying for it so seductively that physicians
build their practice upon it, they squander resources vitally
needed by others. This is an example of the more general problem of medicalizing problems of other sorts, to the detriment of
more useful medicine.

It should be apparent now that my account of qualifications regarding autonomy and intervention has no natural end.
After a more thorough review of the costs of intervention we
would need to weigh various instruments and goals of intervention over against the other values our society prizes: secu-
rity and adventure, just universal equality and attentive partic-
ularity, and so on. But let me stop here and recall one moral
that has emerged from our discussion of principles for
intervening in drug use. When individuals' drug use harms
others, we have little difficulty in saying the others can intervene in their own interests. And when, without harming others, people use drugs with responsible judgment, we readily
accord them the autonomy to do so. But when people's drug
use does not seriously harm others and yet it is not responsible
either, we have no *political* sense of what to do. We fall back
upon arguments about solidarity that define the moral character of society.

A large part of society's dilemma here is philosophic. Liberalism has argued that human reason is basically technological
and analytical—an adequate tool for the pursuit of values, but
not itself the source of values. Liberals have tended to respect
private will as the source of values. But in the competition of
wills, liberalism has retreated to pure procedural prescription.

The above analysis shows that pure proceduralism is not enough; other values are needed to set limits, values that are common and socially warranted. Marxism has accepted the liberal conception of reason as analytic and technological and differs mainly by ascribing the values orienting society to a different source, the dialectic of the control of means of production. Like liberalism, Marxism provides no common social warrant for its values. Conservatives take the radical step of accepting the conception of reason as analytical and technological and then rejecting reason's pretense to justify values. Conservativism falls back on the pure value of tradition and (usually) the status quo. Conservatism turns into imperialism when there is more than one tradition.[8]

The situation calls for a reconstruction of our conception of reason, something inclusive of the scientific integration of qualitative and quantitative thinking but transcending its exclusion of valuational thinking. What shape properly valuational political thinking would take is impossible to say now. Its sources, however, lie in civic republicanism, Confucianism, and pragmatism.[9] At least this is my sense so far. With respect to the birthplace of properly valuational politics, it will not be in theorizing with the categories of liberalism and its opponents, but in reflections on actual valuation in cases such as that of drug abuse.[10]

Notes and References

[1]For background and justification of the first principle see John Stuart Mill's *On Liberty*, New York, Library of Liberal Arts, 1956. For the second principle see John Dewey's *The Public and Its Problems*, New York, Holt, 1927. For Dewey's criticism of the first principle in terms of the second see his *Liberalism and Social Action*, New York, Capricorn, 1963. For a systematic statement of my own development and criticism of Dewey's political theory, see my *Cosmology of Freedom*, New Haven, Yale University Press, 1974, which discusses participation and rights.

[2]*Kaimowitz v. Department of Mental Health*, Civil No. 73-19434-AW (Cir. Ct., Wayne County, Mich., 1973). This case is reviewed at length by Ronald S. Gass in "Kaimowitz v. Department of Mental Health: The Detroit Psychosurgery Case," in *Operating on the Mind*, Willard M. Gaylin, Joel S. Meister, and Robert C. Neville, eds., New York, Basic, 1975.

[3]I have defended the social importance of individual responsibility in "On the National Commission: A Puritan Critique of Consensus Ethics," in *Hastings Center Report*, 9/2 (April, 1979), pp. 22–27.

[4]For arguments about public obligation to develop personal responsibility in other areas, see my "Sterilization of the Mildly Mentally Retarded without Their Consent: The Philosophical Arguments," in *Mental Retardation and Sterilization*, Ruth Macklin and Willard Gaylin eds., New York, Plenum, 1981, pp. 181–193, and "Drugs and Behavior Control: A Dilemma for the Patient, the Nurse and the Physician," coauthored by Jay Schulkin, in *Moral Problems of the Nurse-Patient Relationship*, Catherine Murphy and Howard Hunter, eds., New York, Allyn & Bacon, 1983.

[5]See Peter Laurie's study of drugs in England, *Drugs: Medical, Psychological and Social Facts*, Harmondsworth, Middlesex, Penguin Books, 1967.

[6]For a discussion of pharmacological Calvinists and others distinguished by ideological attitudes toward drugs, see my "Drug Use, Abuse, and Dependence," in *The Encyclopedia of Bioethics*, Warren T. Reich, ed., New York, Free Pres, 1978. The phrase "Pharmacological Calvinism" comes from Gerald Klerman.

[7]On addiction, see Laurie in note 5 above, Dorothy Nelkin, *Methadone Maintenance: A Technological Fix*, New York, Braziller, 1973, and Alfred R. Lindesmith, *The Addict and the Law*, New York, Vintage, 1965.

[8]This point is nicely elaborated by William M. Sullivan in his recent *Reconstructing Public Philosophy*, Berkeley, University of California Press, 1982.

[9]Sullivan makes the case for civic republicanism in *ibid*. Anthony Cua makes the case for Confucianism in his *The Unity of Knowledge and Action*, Honolulu, The University Press of Hawaii, 1982. My *Cosmology of Freedom* argues the case for pragmatism.

[10]The point of this paragraph is developed at length in my *Reconstruction of Thinking*, Albany, State University of New York Press, 1981.

Part II

Pleasure and Performance

The Use of Drugs for Pleasure

Some Philosophical Issues

Dan W. Brock

Introduction

I intend to outline here some philosophical theories of the *nature* of pleasure, and in turn some philosophical arguments concerning the *value* of pleasure, in the hope that doing so will aid in the assessment of attitudes towards the use of drugs to produce or enhance pleasure. In particular, a significant source of the support for continuing strong legal prohibitions on the use of drugs merely for pleasure is the common public attitude that such use is bad. Is there any sound basis for this disapproval of pleasure that is produced by drug use, and if so what can that basis be? And more specifically, is there any sound basis for the disapproval of the use of drugs for pleasure *in itself*, as opposed to disapproval because of other consequences such use has, or may be thought to have?

Although I shall be examining philosophical issues and arguments here, I believe it would be a mistake to think that popular attitudes toward the use of drugs for pleasure rest largely on any well-founded and thought-out philosophical position. As some of the other writers here make clear, many historical and political factors have contributed to present attitudes toward the use of drugs for pleasure in ways that often do not stand up to rational scrutiny. It would, consequently, be a further mistake to suppose that philosophical confusions are the principal source of more general confusions and irrationalities in attitudes towards the use of drugs for pleasure, and in turn that removal of this philosophical confusion would clear up most of the more general confusion and irrationality found in the area. In particular, and as I shall elaborate briefly at the end of the article, most such confusion is probably of an empirical rather than philosophical sort—confusion in the form of myths and other unfounded or false beliefs about the nature of these drugs and the various circumstances and consequences of their use. There may, nevertheless, be room for some illumination of the area if we can clear up some of the philosophical confusions that have a part in attitudes about the use of drugs for pleasure.

I should note that the focus of my concern is limited in another way. I shall not discuss the general issue of when the state may rightly regulate or prohibit behavior, the different justifications for its doing so, nor, more specifically, which if any of these justifications may apply to the use of drugs for pleasure. These are obviously important issues of moral and political philosophy for the larger project addressed in this volume. My concern then will be the modest and limited one of attempting to gain some philosophical clarity about the value or disvalue, goodness or badness, of the use of drugs for pleasure.

When we consider opposition to the use of drugs for pleasure, there are two obvious sorts of objections that we might expect to find. First, it may be that *pleasure* is what is opposed. It is the goodness of pleasure, or its place in a sound account of the good for persons that is being questioned. But that is unlikely since, except perhaps for extreme ascetics, few persons oppose pleasure generally, or all pleasure. Second, it might be the use of *drugs* that is opposed, but that too would be problematic since the use of drugs for other purposes, such as to alleviate pain, is rarely opposed. So it must be the particular combination of the use of *drugs* in order to produce *pleasure* that is opposed. To evaluate whether there is any sound philosophical basis for this opposition, we need to consider several issues. First, we need to get clearer about the concept of pleasure, what pleasure is. Having done that, we need then to consider what value is to be assigned to pleasure, and whether and if so, how that value may vary according to the source of the pleasure. Finally, we will then need to consider the place of pleasure, and then specifically the use of drugs for pleasure, in a conception of the good for persons.

The Nature of Pleasure

First, then, the nature of pleasure. It may be that we all know pleasure, and even more clearly pain, when we experience it, but to give an account of what pleasure is—an analysis or definition of the concept of pleasure—is no easy task. There are a variety of experiences of pleasure or enjoyment that any analysis of pleasure should cover. For example, there is the pleasure from sexual orgasm or a good meal, as well as the rather different enjoyment from reading a good book or spending a day in the country. These two different classes of cases each in turn lend plausibility to two major alternative philosophical theories about the nature of pleasure. The first theory I shall call the property of conscious experience theory (PCET). On this account pleasure is a sensation, feeling, or quality of conscious experience that is liked or found agreeable for its own sake.[1] Thus, a back rub, sexual orgasm, or smoking a marijuana cigaret may produce pleasurable feelings or sensations. It is on this sensation or feeling sense of pleasure that pain is naturally taken to be the opposite of pleasure, a sensation or feeling disliked or found disagreeable for its own sake. It is probably this sense of pleasure that people usually have in mind when they think of drugs used for pleasure, that is, drugs used to produce particular pleasurable feelings or sensations, or to enhance the pleasure from already present feelings or sensations.

It is on this first view of the nature of pleasure that it may seem natural to think that activities are undertaken *for* the pleasurable feelings or sensations they produce, and in turn to think that, were it possible to have those same pleasurable feelings and sensations without having to undertake the activities that produce them, then we would do just as well without the activities. The hedonist, who attributes value to all pleasure, and only to pleasure, is on this account of pleasure correctly viewed as attaching no intrinsic value to the activities that persons normally pursue and that produce pleasure. On this account of pleasure, pleasure seems to have what might be called a detachability from the activities that produce it. It is the pleasurable sensations or feelings that are liked for their own sake, and the activities that produce the pleasure only happen to be contingently necessary to the pleasure. A hedonist's life filled with pleasure on this view could as well merely contain the feelings or sensations without the activities, if that turned out to be empirically possible.

Enjoying or taking pleasure in reading a book or spending a day in the country, however, seem not to fit the sensation or feeling view. Once we look at enjoyable activities, like these, that tend to last a significant period of time and to contain significant variety, it seems difficult to locate any sensation, feeling, or other quality of conscious experience that is the pleasure. We enjoyed the whole book or day in the country, but there was no sensation or feeling that continued throughout each experience that was the pleasure. In general, if we look at the enormous variety and duration of experiences that we enjoy or take pleasure in, there seems no unitary sensation, feeling, or quality of conscious experience that is present throughout them and that can be identified as the pleasure. This has led philosophers to an alternative account of pleasure that I will call the preference theory (PT). On this account, one is enjoying (taking pleasure in) an experience if, at the time, one likes it for its own sake, in the sense of wanting to sustain and repeat it for its felt qualities and apart from its consequences.[2] Pleasure on this second view refers not to a property of conscious experience, but to a relation holding between an experience and a person's preferences or desires. To say that one enjoys x, is to say that one has a non-evaluative, pro-attitude towards x at the time one undergoes x; to say that the attitude is non-evaluative is to say that the positive attitude or preference is not based on any evaluation (moral or otherwise) of the person that he ought to prefer or have a positive attitude towards experiences of this kind.

There are many refinements in either the conscious experience or preference conceptions of pleasure that would be necessary before we could fairly evaluate whether either is a satisfactory account of the nature of pleasure, but I shall not concern myself with most of them since they are not central to my concerns in this paper. But a few points should be noted. First, on the preference theory, since pleasure refers to a relation that holds between a person's preference and his or her experience, there should be less temptation to believe, as I noted above did seem plausible on the conscious experience view, that one could get the pleasure without the experience or activity in which one takes pleasure. As I shall discuss further below, one basis for the objection that a life filled with pleasure is not a good life is that on the conscious experience view that seems consistent with a life without all the various activities with which we in fact seek to fill our lives, so long as we could get the conscious experiences without the activities. The philosopher J. J. C. Smart has imagined being able to hook oneself up to electrodes that would stimulate parts of the brain so as to produce the conscious experience or feeling of pleasure without the usual activities necessary for such feelings.[3] If a life hooked up to such electrodes fails to conform to the conception of the good life that many of us have, as I believe it does, a life filled with pleasure on this conception of pleasure need not be a good life. That sort of objection is not as easily made against a life filled with pleasure on the preference conception, since the activities are logically necessary to the pleasure, given that the pleasure refers to a relation holding between the person's preferences and his activities.

Second, the preference theory allows a hedonist to meet the common objection against hedonism on the conscious experience account that there seem to be many activities that we seek and value that do not produce any identifiable feelings or sensations of pleasure. We have already mentioned two examples, reading a good book and spending a day in the country. Examples could, obviously, be multiplied indefinitely, since on the preference view *any* experience that, at the time one undergoes it, one likes for its own sake—that is, for its own felt qualities and apart from its consequences—is an experience in which one is taking pleasure or enjoying. In particular, all of the so-called "higher" or "distinctly human" activities that persons like for their own sake, such as understanding a philosophical argument, good conversation between friends, and so forth, are experiences in which we take pleasure on the preference view. J. S. Mill considered the objection to hedonism that it was a doctrine fit only for swine, in that it attributed value for humans to only the lowest or basest animal pleasures.[4] And some such view may well underlie the distrust and disapproval with which many Americans regard pleasure and hedonism. Mill rightly noted that this objection was misplaced, in that it rested on a conception of humans as only taking pleasure in the experiences shared with swine, whereas because humans are the kinds of beings that enjoy experiences, of intellectual and other sorts, of which swine are incapable, the hedonist can appropriately ascribe value to our pleasure in these "higher" experiences.

Earlier we noted that disapproval of the use of drugs for pleasure might have its source in general disapproval of pleasure. When it does, it often takes some form of the "swine objection" above, that is, that persons are capable of many higher activities, and to view persons as merely pleasure-seeking beings is to debase their nature. But we can see now that such an objection rests on the same confusion as did the swine objection, and is to be answered in the same manner as Mill answered it. Just in case human beings are capable of, *as they are*, and in turn take pleasure in, *as they do*, "higher" activities, then disapproval of pleasure is not justified by a belief that it is only "lower" activities that lead to, or are associated with, pleasure. It is worth adding that even if it were true that only "lower" activities lead to pleasure, this would seem to be a reason to value other things ("higher" activities) besides pleasure, but not a reason to assign disvalue to pleasure.

With these two broad philosophical accounts of the concept, and nature, of pleasure before us, it is worth briefly asking how these philosophical theories are related to recent neurochemical advances concerning pleasure, and in particular the discovery of endorphins. As discussed elsewhere in this volume, recent research has shown that the body contains specific receptors for the opiates, such as morphine. This suggested that the artificial opiates must have their analogs in natural substances, the endorphins, that are "manufactured by" or at least present in the body naturally and that bind to these receptors. Perhaps then these natural opiates are the more general pleasure producers/pain inhibitors in the body, just as morphine may be an externally induced artificial pleasure producer/pain inhibitor. Suppose future research allowed us to isolate specific neurochemical changes that occur in the brain every time a person undergoes a pleasant or unpleasant experience, every time he/she has a pleasure or a pain. The implications of such a finding would likely be many, but what specifically would they be for the philosopher's concept of pleasure? And that means as well, for the ordinary person's concept of pleasure, since the philosopher's concept is intended to be an analysis of it. Would it, for example, be the case that we would then know what pleasure *really* is, and could dispense with the

philosopher's and ordinary man's prescientific account? I
would suggest not. What we would be faced with would be a
specific instance of what has traditionally and more generally
been called in philosophy the mind–body problem.[5] The philosopher's and ordinary person's concept of pleasure concerns
their conscious experience, referring either to a property of a
person's conscious experience or to a relation between this conscious experience and one's preferences. The "new conception
of pleasure" would concern our physical states, and be formulated in physical, specifically neurochemical terms. The two
conceptions operate at two different levels of discourse and of
reality, one at the mental level of the conscious experience of
persons, the other at the level of physical states or organisms.
The philosophical problem would then be how the two, the
mental and physical, are related. On one view, usually called
materialism or physicalism, the relation is one of identity with
the mental being in some sense reducible to and nothing more
than the physical.[6] But this is only one position, and a highly
controversial one, concerning the nature of the relation. The issues and literature concerning the mind–body problem are
enormously complex, and need not detain us here. The point
to be stressed is that advances in our physical understanding of
humans do not uncontroversially give us reason to give up the
language of the mental in which the concept of pleasure is
found. Rather, it would be more accurate to say that such advances increase our understanding of what physical, and specifically neurochemical, changes take place in the body when a
person experiences pleasure. More specifically, we may be able
to compare what differences there are between the neurochemical effects accompanying pleasurable experiences of an
ordinary, "natural" sort, as opposed to pleasure induced by ingestion of artificial opiates. If the latter have some harmful consequences, for example, in the developmental process of young
adolescents because of the large doses of endorphins introduced, that may give us good reason to be cautious about uses
of opiates in those cases. None of this, however, would establish that the philosophical accounts of pleasure are mistaken, or
that attitudes toward pleasure itself, as opposed to different
sources of pleasure, are in need of revision.

     Given the two philosophical theories of pleasure that we
have noted, PCET and PT, which one is correct? As I have
noted earlier, attempting to answer that question would re-

quire us to introduce many refinements into the two theories, and to assess many issues and arguments, most of which would not serve my purposes in this paper. So we must be content here to leave that question unanswered, but we can consider the more restricted question of which theory is appropriate when we consider the use of drugs for pleasure. The narrower, property-of-conscious-experience theory, which holds that pleasure is a sensation, feeling, or other property of conscious experience, seems to fit many uses. Descriptions of the pleasurable effects of drug use by drug users are often in terms of sensations or feelings that the drugs produce. On the other hand, sometimes the descriptions do not focus on unique sensations or feelings from the drugs, but rather on the way in which the drugs enhance the pleasure obtained from experiences that were pleasurable independent of the drug use. For example, the effects of marijuana in focusing one's attention and shutting out distracting thoughts or sensations in order to enhance the pleasure of sexual activity or listening to music is of this sort. The pleasure is still in the experience of sex or listening to music, but is intensified and focused by the drug effect. The drug is a pleasure enhancer, rather than a pleasure producer. And that may be true even for "long term" experiences that seem best to fit the preference theory, for example, "being stoned" during the day in the country. It is thus unclear that only one of these two theories provides the most plausible analysis of pleasure in all its "drug use" instances, much less in all instances, and both theories must here be left in contention for the position of correct theory.

Higher and Lower Pleasures

I shall turn now to some of the normative or moral issues that we need to address here. These can be formulated in a number of ways, keeping in mind the aim of attempting to clarify whether there is any philosophical basis to the opposition to the use of drugs for pleasure. Few of the opponents of the use of drugs for pleasure oppose all pleasure, so one issue can be put as, is there any basis for evaluating some pleasures as better than others, such that pleasure from drugs might come out lower on the value scale? Mill, for example, made a well-known distinction between higher and lower pleasures, higher and lower in an evaluative sense, which would seem relevant here. And a second question, which I shall take up in the next section, is what place does pleasure have in a sound theory of the good for persons, what is the proper role of pleasure in the good life? I believe these questions turn out to have a different form and meaning depending on which of the two accounts of pleasure we are using, and that the plausibility of different answers to them depends in part on how pleasure is understood.

Before directly addressing whether pleasure from drug use can plausibly be taken to be a lower pleasure, we need to bring out more explicitly a difference in the two theories of pleasure that was implicit in some of our discussion of them, a difference in the components of their respective analyses of pleasure. On the property of conscious experience theory (PCET) there are the following components:

1. The pleasure, which is a sensation, feeling, or other property of conscious experience.
2. The conscious experience of the person, of which the pleasure is a part or aspect.
3. The source of the pleasure.

The source of the pleasure will be some activity or behavior, for example playing a game of tennis or smoking a marijuana cigaret, that produces component 2 above, the conscious experience, and component 1 above, the pleasure that is a part or aspect of that experience. On the preference theory (PT) there are these components:

1. The person's experience, which can include two components:
   a. A conscious experience, e.g., thinking thoughts, feeling hot or tired, and so on.
   b. An activity, e.g., playing tennis, having sexual intercourse, reading a book, and so on.
2. The person's liking the experience, in the sense of desiring or preferring at the time of undergoing it to continue and repeat it because of its felt qualities and apart from its consequences.
3. The pleasure, which just is component 2 holding of component 1, that is a positive desire or attitude towards the experience.

On this second theory, pleasure is nothing over and above the relation between a person's experience and likings, where the likings are analyzed in terms of the person's desires.

Now an *extreme opponent* of the use of drugs for pleasure would be one who denied that pleasure having its source in drug use was good *at all*, or had any intrinsic value. He or she would be denying the first of the following two claims, which I have noted together define ethical hedonism:

1. All pleasure has intrinsic value (is good, desirable).
2. Only pleasure has intrinsic value (is good, desirable).[7]

The denial of claim 1 would rest, at least in part, on the example of drug use. But it is very difficult to see what *argument* the extreme opponent could offer to establish that pleasure having its source in drug use has *no intrinsic value at all*. This difficulty can perhaps best be brought out by briefly considering one kind of case that *has* been taken to present a strong objection to the first claim of hedonism, and how the hedonist may answer that objection.

Consider malicious or sadistic pleasures, cases where one person's pleasure is in the misfortune or suffering of others. In general, philosophers have argued that pleasure always takes an object, pleasure must be *in* or *with* something, and here the object is another's suffering. The usual response to the objection that there is *no* value in sadistic or malicious pleasures turns on the distinction between intrinsic and extrinsic value. Roughly, that distinction is between the value a state of affairs or experience has in itself, or because of its own nature, as opposed to the value it may have on account of its consequences or relations to other things. We might then say that sadistic or malicious pleasures will always have the *consequence* of the suffering or misfortune of another, and in general the latter might be thought likely to be greater than the former. This line of reasoning is a way of explaining one's opposition, all things considered, to sadistic or malicious pleasures, without abandoning the ethical hedonist's position that all pleasure has some intrinsic value. On PCET, where pleasure is the logically detachable component 1 noted above, this line of response seems plausible. On PT, where pleasure is a relation between components 2 and 1, and where 1 includes an activity, one might think that the causing of suffering of another could not be separated from, and so merely an extrinsic consequence of, the pleasure. But such pleasures need only be at the *thought* of the suffering or misfortune of another, and need not involve one's doing anything to cause that suffering. And, perhaps more important, since such pleasures require only the belief that others are suffering, that belief could be false and so the sadistic pleasure occur without anyone else suffering. J. J. C. Smart has asked us to

> . . . *imagine a universe consisting of one sentient being only, who falsely believes that there are other sentient beings and that they are undergoing exquisite torment. So far from being distressed by the thought, he takes a great delight in these imagined sufferings. Is this better or worse than a universe containing no sentient beings at all? Is it worse, again, than a universe containing only one sentient being with the same beliefs as before but who sorrows at the imagined tortures of his fellow creatures?*[8]

Smart's example is designed to focus more precisely the question of whether the sadistic pleasure, in itself and apart from its consequences, is bad, and I believe he is correct in suggesting that it is not. But if the distinction between intrinsic and extrinsic value, coupled with the method of analysis employed in Smart's example, are sufficient to defend the intrinsic value of even sadistic pleasures, or at least to undermine the force of sadistic pleasures as objections to the first claim of hedonism, then they will probably be sufficient against any argument the extreme opponent of the use of drugs for pleasure might offer. It is highly unlikely that the extreme opponent of the use of drugs for pleasure can make out a plausible case that pleasure from the use of drugs either has *no* positive intrinsic value *at all*, or has intrinsic *dis*value.

A more *moderate opponent* of the use of drugs for pleasure is one who does not deny the first of the two theses of ethical hedonism (or at least does not deny it for the case of drug use), but rather claims that some pleasures are better or more valuable than others, and that pleasure from the use of drugs is a lower value or quality pleasure. As already mentioned, J. S. Mill made a well-known distinction between the qualities, as well as quantities, of pleasures, though he did not apply the quality distinction to the case of drug use.[9] All hedonists have agreed that pleasures can differ in quantity, in particular in amount at a point in time, and in duration over time. Mill's claim, and the claim of our moderate opponent of drugs used for pleasure, is that pleasures can differ in quality as well as in quantity, where "quality" denotes evaluative differences, not merely differences in kind that do not mark nonquantitative evaluative differences. On this view, though pleasure A might be of greater quantity than pleasure B, B might be of higher quality, and so of overall higher value than A. Mill's criterion for which pleasures were of higher or lower quality was to appeal to the preferences of those who had experienced both sorts of pleasures; where such qualified judges preferred B to A, though A was of equal or greater quantity than B, then B was a higher quality pleasure than A. There are many difficulties of both a practical and theoretical sort with such a proposal, and I shall mention only those that are important for the position of our moderate opponent of the use of drugs for pleasure, and will illuminate the core of plausibility in the quality of pleasure distinction.

A common objection to Mill's introduction of a distinction between qualities of pleasure is that it was inconsistent with his acceptance of ethical hedonism and its two theses noted above. Now since it is no part of my aim in this paper to defend ethical hedonism, we might simply conclude that if the "quality of pleasure" distinction is sound, then so much the worse for ethical hedonism. But both what makes appeal to some "quality of pleasure" distinction plausible for a hedonist, as well as what generates the supposed inconsistency, *are* important for our moderate opponent's position, and so we must look more carefully at the difficulty here. The supposed inconsistency can be brought out most clearly on the PCET account of pleasure. On that account, pleasure is a sensation, feeling, or other quality of conscious experience. And it is usually understood to be a single, unitary sensation, feeling, or property of conscious experience, single or unitary because it is an unanalyzable quality, much as color qualities of experience, such as redness, are taken to be; pleasure is a property of our conscious experience that cannot be broken down in analysis into other properties. The hedonist claims that it is the presence of this, and only this, unanalyzable property of pleasure in an experience or state of affairs that gives value to the experience or state of affairs. But to claim that pleasures differ in quality (evaluative sense) as well as quantity, seems to be to require some *other* property besides pleasure to be the *basis* for the difference in value, and so which itself gives value to states of affairs or experiences. But this seems inconsistent with the hedonist's claim that *only* the property of pleasure gives value to states of affairs or experiences.

Moreover, apart from this inconsistency with hedonism, on PCET it is difficult to see how pleasures in themselves could

differ qualitatively as Mill's distinction supposes, if pleasure itself is in fact a basic, unanalyzable property of our conscious
experience. We can see easily enough how components 2 and 3
of the PCET analysis above could differ qualitatively, but they
are not, on this analysis, strictly speaking the pleasure. Thus,
on PCET, the position of the moderate opponent of the use of
drugs for pleasure is not merely inconsistent with hedonism, in
which case one might simply give up hedonism, but is thoroughly obscure concerning how, on that view of the nature of
pleasure, there could possibly *be* higher and lower pleasures,
pleasures whose differences in intrinsic value were of a qualitative and not merely quantitative sort. This strongly suggests
that, on the PCET account of the nature of pleasure, differences
in the overall value of experiences or states of affairs that are
not based on differences in quantity of pleasure in that state of
affairs will be differences in the extrinsic, not intrinsic, value of
those states of affairs. And this means that on the PCET account of the nature of pleasure, pleasure from the use of drugs
is not in itself a lower quality or value pleasure than pleasure
from any other source, but is of lower value overall (if it is) only
because of its consequences or, as it might be put, side effects.
The defense of this latter position would require appeal to empirical claims about the nature of the consequences or side effects of the use of drugs for pleasure, and I shall say something
about such claims later in Section III, but it is not an explication
of the position of the moderate opponent as we have understood him here.

If we turn now to the PT account of pleasure, we can perhaps make more sense of the quality of pleasure distinction,
and bring out more clearly what makes the distinction plausible. On this view, recall, to say that a person is taking pleasure
in an experience just is to say that he is undergoing that experience, with its conscious and activity components noted above,
and that he likes that experience, in the sense of desires to continue and repeat it, for its own sake and apart from its consequences. The pleasure consists in the relation holding between
his experience and his liking, "liking" analyzed in terms of his
desires or preferences. One of the elements of this relation, his
experience, with its two components of his conscious experience and his activity, is clearly such that it can vary in quality in
a multitude of ways that can affect the pleasure relation by affecting his liking of the experience. For example, compare two

sexual experiences: masturbation and sexual intercourse. Although the sensation of sexual orgasm might be the same with each, because the latter can involve emotions, feelings, beliefs, commitments, and so forth, that the former does not, we can speak of the two varying in quality in an evaluative sense, and of sexual intercourse being the richer and higher quality experience. But is it a higher quality pleasure, independent of differences in quantity of pleasure? There are two interpretations open to us here. If the quantity of pleasure is determined by the degree of liking of the experience, which is in turn determined by the strength of desire to continue and repeat it, and if the differences in kinds of experience merely contribute to differences in degree of liking of them, then it seems we may still have here only a difference in quantity of pleasure. On the other hand, perhaps further elaboration of this theory of pleasure would show this to be a difference of quality rather than quantity. But on either interpretation, the important point is that the PT account of pleasure can accommodate the underlying intuition that supports the plausibility of distinguishing qualities of pleasure in the first place, namely that in the actual valuing that persons perform, many features or properties of an experience contribute to our overall evaluation of it. And this, I believe, is a crucial advantage of the PT account for a defender of hedonism. In general, objections to the second thesis of ethical hedonism, that only pleasure has intrinsic value, take the form of arguments that many things besides pleasure have intrinsic value. Persons value friendship, commitments, the respect of others, making their own choices, and many other features or properties of experiences or states of affairs, besides pleasure. Ethical hedonism on the PCET account of pleasure seems to fail to attribute intrinsic value to these features of experience, and to be an inadequate theory of value for that reason. Ethical hedonism on the PT account of pleasure does reflect this diversity or plurality in most persons' actual values. The PT account of pleasure makes sensible the evaluation both of some pleasures as better than others, and of some sources of pleasure as better than others. Some pleasures can be better than others, whether this is, in the end, in a qualitative or quantitative sense, because the experience that the pleasure relation holds with regard to, contains properties or features to which is ascribed independent, nonderivative value. And some sources of pleasure can be better than others in the very same way—the different experiences that are the sources of our pleasures, that is the causes of our liking the experiences in varying degrees, can themselves contain other valued, that is liked, elements. If the above is sound, it explains how it can be sensible to talk of some pleasures, and of some sources of pleasure, being better than others. But the moderate opponent of the use of drugs for pleasure, while of course requiring that such talk be sensible, makes the additional specific claim that pleasure having its source in drug use is among the lower quality pleasures. What particular features of drug-induced pleasure may underlie the moderate opponent's view, I shall suggest later in the next section, but the general structure of his view can be noted now. Our earlier example of sexual activities is instructive. Masturbation is thought to be a lower quality source of pleasure than sexual intercourse because it does not contain elements of love, commitment, and so forth, that sexual intercourse can contain, and that can contribute (whether finally, in a qualitative or quantitative way) to the latter's value. Sexual intercourse is "value rich," as we might put it, in a way that masturbation is not. Likewise, the moderate opponent of the use of drugs for pleasure will hold that drug-induced pleasures are "value poor" in comparison with many other common pleasures, in that they lack valued features or properties of experience that such other pleasures contain. Again, I shall suggest what some of these valued features are in the next section. But it should be added here that the moderate opponent of the use of drugs for pleasure often holds not merely that such pleasure is of a lower quality kind, but that it tends to replace pleasures of a higher quality kind in the lives of persons who resort to it. This too, of course, is an empirical claim for which evidence is needed. Let me emphasize that I am not arguing that pleasure from the use of drugs in fact *is* a lower quality pleasure, but am only explicating how we can make sense of such a view. I shall note in the next section some of the empirical claims that would have to be supported if one wanted as well to defend the moderate opponent's view. But it should be stressed that even if the pleasure from drug use is a lower quality pleasure, it is still a pleasure and so, even on the moderate opponent's view, an intrinsic good, albeit a lesser one.

The Place of Pleasure from Drugs in the Good Life

We must now try to clarify the second major normative or moral issue that I noted earlier in this paper—what place does pleasure, and then more specifically the use of drugs for pleasure, have in a sound theory of the good for persons, or what is their proper role in the good life? The problem with these questions posed in this way (and of course I am responsible for posing them in this way) is that, certainly on the PT account of pleasure, they are misleading. Asking what place or role pleasure has in the good life seems to assume that pleasure is one goal or end among others that we seek in our action, and the question is then how much pleasure, as opposed to other things, we should seek. This question about pleasure seems analogous to such questions as what is the proper place or role of playing tennis or playing with one's children in the good life. How much time should I devote to tennis or my children, and likewise how much should I devote to pleasure. But clearly on the preference theory of pleasure this is a confusion. Pleasure is a relation between my experience and my desire to continue and repeat the experience for its felt qualities and apart from its consequences. The goal or end is the experience, for example, playing tennis or playing with my children, and the pleasure is not one experience, and so one goal among others to which we must assign proper weight, as our question above suggested. I shall try shortly to give some sense to this question, but whatever sense it has is not to be found along these lines.

There is a related confusion that should be removed before we directly address the place of the use of drugs for pleasure in the good life. I have spoken often in this paper of the use of drugs for pleasure, and the implicit comparison is with their use for other purposes, specifically therapeutic purposes in the treatment of illness and disease. This comparison too can misleadingly suggest that some activities of life are aimed at pleasure, whereas other activities are aimed at something(s) else, and the problem of the proper role or place of pleasure in the aims of a good life again arises. But this is not a proper understanding of what is usually meant by the distinction between doing something for pleasure as opposed to for some other end. Consider two tennis players. Player A plays tennis for pleasure, which on the PT account of pleasure just means that at the time A plays tennis he or she likes it, in the sense of desiring to continue and repeat the experience of playing tennis for its own sake and apart from its consequences. Player B plays tennis not for pleasure—in fact B rather dislikes playing—but for the entree it provides to otherwise unavailable social circles and activities that B enjoys. It is not as if pleasure has a greater place in A's life than in B's. Rather, playing tennis is a direct source, is itself a source of pleasure for A, whereas for B it is not a direct source, or itself a source of pleasure, but is only instrumental as a means to something else, and only that something else, the social activities, is itself a source of pleasure for B. This is a distinction between an experience desired or enjoyed for its own sake, as opposed to desired as a means to some other experience that is desired or enjoyed for its own sake. In each case there is an experience desired or enjoyed for its own sake,an experience in which the person takes pleasure, and so pleasure does not have a more prominent place in A's life, and in particular in A's scheme of ends, than in B's. If one is to talk of using drugs for pleasure, the less misleading comparison is with the use of other things for pleasure, that is, with other experiences also liked for their own sakes, rather than with the use of drugs for other things or other ends instead of pleasure.

The above is not to say, however, that any experience desired for its own sake must be an instance of pleasure. The preference theory's analysis of pleasure in terms of desire—specifically the desire to continue and repeat an experience at the time one undergoes it, and for its felt qualities and apart from its consequences—commits the preference theory to all pleasure being desired. But it does not commit the preference theory to *only* pleasure being desired, as an analysis of desire in terms of pleasure might do. It is consistent with the preference theory's analysis of pleasure that one desire something that will not produce pleasure. I can today desire to have an experience tomorrow that I now correctly believe I will not like, that is, not desire to continue and repeat for its felt qualities when I have the experience tomorrow. For example, I may believe that it is my duty to have the experience, though I know I shall find it distasteful, and more generally, I may desire other things besides pleasure because I can have other reasons to desire to have an experience besides the reason that I will like its felt qualities when I have it. When other things besides pleasurable experiences are desired for their own sakes, the idea of pleasure as one end among others is sensible.

What can we now say of the place of pleasure whose source is drug use in the good for persons or the good life? One difficulty, as we shall see, in attempting to provide an answer to this question is that the major alternative philosophical theories of the good for persons are at too general or abstract a level to have much in the way of clear implications on this question. The differences between these theories concern theoretical issues that are at a considerable remove from the question of drugs used for pleasure. And within any of the major theories considerable empirical data, largely concerning the effects of drug use, will be necessary before any of the theories would have any clear implications concerning the use of drugs for pleasure. I believe that most of the disagreement concerning the desirability of the use of drugs for pleasure rests not on philosophical, and more specifically moral, differences concerning the good for persons, but rather on the dispute over the empirical facts concerning the consequences of the use of drugs for pleasure. In the remainder of this paper, I shall elaborate these points and suggest where I believe the controversial issues do lie.

What are the major alternative philosophical theories of the good for persons? I shall distinguish three: the pleasure or happiness theory; the desire theory; the perfectionist theory. It is common, and for historical accuracy within the history of philosophy, necessary to distinguish between pleasure and happiness, and in turn between pleasure and happiness theories of the good. I shall not do so, however, because they are similar, if not identical, in the respects in which I am now interested. Both can be construed along the lines of the account of pleasure in the PT theory. To promote a person's good on the pleasure or happiness theory is to promote one's having experiences that, at the time they are undergone, are liked for their own felt qualities.[10] When having such experiences, one experiences pleasure, enjoys oneself, or is happy, and the more one likes the experiences, the more pleasure one has or the happier one is. In a crude nutshell, on this first theory, a person should attempt to have as much experience that pleases or makes one happy, when he or she has it, as possible.

On the desire theory, what is good for a person is for his or her desires to be satisfied to the maximum extent possible.[11] It might seem that since we have construed pleasure, on the preference theory, in terms of desiring to continue and repeat an experience, that there is no substantial difference between a pleasure/happiness theory of the good and a desire theory. However, that is not the case. As noted above the preference theory is committed to all pleasure being desired for its own sake, but not to only pleasure being desired for its own sake. The difference between the pleasure/happiness and desire theories of the good for persons is perhaps most obvious in the case of desires whose objects are not one's own conscious experiencings, e.g., my desire that my children thrive, and thrive not just while I am alive to take pleasure in their thriving, but after I die as well. This desire is satisfied by it being true that my children do thrive, both before and after my death, though in the latter case I will not be around to receive satisfaction from their thriving. More importantly for our purposes, it seems possible that a person desire something even though he or she, correctly, does not expect it to give pleasure or make one happy, either at all, or as much as something else that one does not desire or desires less. Although the desire theory enjoys considerable popularity among contemporary moral philosophers, and in fact has important attractions, serious difficulties have been raised concerning whether it is even a coherent, much less a morally attractive, conception of a person's good.[12] Moreover, there are significantly different versions of the desire theory; there are various constraints one can plausibly put on which desires are such that their satisfaction is for a person's good, e.g., only the desires that have a place within a person's rational plan of life. But I shall leave all these difficulties and details aside here. The main point of contrast is that *if* we want other things besides what will make us happy, or want things that will not make us happy or give us pleasure, the desire theory takes account of this in its account of a person's good.

The third sort of theory of the good for persons is a perfectionist or ideal theory. On this view, some experiences or states of affairs may have a place in a person's good at least in part independent of any pleasure or happiness they will produce for that person, or of whether they are desired by that person.[13] Perfectionist theories will vary according to which particular ideals are defended, and how the ideals are argued for or defended. Examples of such ideals include developing one's special talents, serving God's will, being an autonomous chooser, being nice to people you don't like and so forth. It is to be emphasized that perfectionist theories generally allow that happiness and/or desire satisfaction are *part* of our good, and the ideals then either constrain the production of happiness or desire satisfaction, *and/or* serve as additional elements of our good.

The above is only the merest indication of the general form of these three sorts of theories of the good for persons. But to know the place of drug use in the good for persons on any of these theories of the good for persons, it is clear that we also need considerable empirical information of two sorts: (1) information about persons, in particular about what makes them happy, or what they desire; and (2) information about drugs, in particular about the various effects that use of drugs for pleasure has. Without such data, the different theories of the good for persons are simply at too high a level of generality to have definite implications concerning the use of drugs for pleasure. But even without such information, we can note that there is a *prima facie* case that the use of drugs for pleasure is part of some person's good on all three sorts of theories. This is most obvious on the pleasure/happiness theory, since persons who take drugs because they in fact produce pleasure are producing what that theory takes to be good. On desire theories, when a person freely takes the drug because he or she desires the experience it produces, its use is again *prima facie* for the person's good. On perfectionist theories, its use may be part of a person's good if enjoyed or desired for the same reason as above. Perfectionist theories would differ from pleasure/happiness and desire theories only if they contained a specific ideal according to which the use of drugs for pleasure was *in itself* either intrinsically good or bad. So far as I can tell, no plausible perfectionist theory has contained such an ideal.

The issue would then appear to be what, if anything, might defeat the *prima facie* case that the use of drugs for pleasure is a part of the good of those persons who do in fact enjoy and/or desire their use? I would suggest that any rational basis for the position of the critic of such drug use must lie in arguments that the consequences of the use of drugs for pleasure are such that they interfere with or inhibit realization of other important aspects of a person's good. Such arguments must include two different sorts of claims: first, claims that particular things are, on some general theory of the good for persons of the sort noted above, components of a person's good; second, empirical claims that use of drugs for pleasure has the consequence of inhibiting or interfering with the promotion and realization of these other aspects of a person's good. If we move down one step from the high level of generality of theories of the good for persons noted above, I believe we would find fairly widespread agreement among proponents of all three theories that the following are components of a person's good. I shall very briefly state these components, but not argue for their being part of person's good, though I believe such arguments could (at some length) successfully be made; the sort of argument necessary would vary according to which general theory of the good is adopted. All of the following then, I believe, are part of the good for persons on any of the usual theories of the good for persons:

1. Participation in social relations of friendship, love, and intimacy.
2. Development and exercise of certain distinctive human capacities, especially intellectual capacities.
3. Being responsible in one's relations to other persons.
4. Being an active person, actively participating in a full array of activities and experiences.
5. Being autonomous, at least in the sense of having and exercising the capacity to act in the manner that on reflection one values or most wants to act.

I shall not elaborate the precise meaning of any of these components of a person's good, but shall rely on the reader's loose understanding of them, and agreement that they *are* part of our good. The point is that there is a widespread popular image of the user of drugs for pleasure, on which the realization of each of these components of our good is interfered with by such drug use. This popular image is clearest probably concerning the heroin user, but I believe it exists as well in a somewhat attenuated form for users of drugs like cocaine and marijuana. On this image, the drug user is:

1. Withdrawn from others, existing in his or her own "private world," and so having, at best, impoverished social relations with others.

2. Dominated or consumed by drug use, in a way that
   precludes the development and exercise of one's other
   human capacities, including especially intellectual
   capacities.
3. Irresponsible in one's relations with others. This is because of both the extent to which drugs are thought to
   dominate or "take over" the user's life (2 above) and to
   produce strong, irresistible desires for the drugs ("the
   heroin user would sell his sister to get money to feed
   his habit").
4. Passive, "doped up," "nodding out," so that one is
   unable to actively engage in a rich array of life's experiences (recall in *Brave New World* how Soma was used to
   produce passive persons who avoided many of the experiences most persons value).
5. Addicted, unable to resist the desire for the drug even
   when one wants to do so, and so under the control of
   the drug as to be the paradigm of the nonautonomous
   person.

I have deliberately used some of the vague and evaluatively
loaded language of this popular image of the user of drugs for
pleasure. But if this popular image is accurate, then the objection to such drug use would, I believe, be sound, for the use of
such drugs would interfere with important components of our
good. If this popular image is true, we would then have a
sound basis for the view that pleasure from the use of drugs is a
lower quality pleasure than other pleasurable experiences that
either themselves promote, or at least do not interfere with, the
five components of person's good noted above. The controversial point is not likely to be whether the five components I have
cited are indeed part of the good for persons, but whether
drugs (or some drugs, in some uses) do in fact prevent their
realization in the way that what I have called the popular image
suggests. That can only be determined by empirical studies,
not philosophical analysis, and other contributors to this
volume are in a better position than I to assess the evidence
concerning the consequences of the use of drugs for pleasure;
the relevant data is both substantially incomplete and contro-
verted. But almost certainly, I believe, such an assessment
would show that the popular image is vastly oversimplified
and substantially, probably largely, false. If that is so, then
what is wrong with the popular image is not its implicit philosophical basis, specifically the components of the good for per-

sons that I have noted above and that it implicitly adopts, for those *are* part of a sound account of the good for persons. What *is* wrong is the false empirical claims or assumptions it contains concerning the effects of pleasure producing drugs on their users. It is the task of empirical research, and of social and natural scientists, not of philosophers, to establish where these claims are false and where they have some evidentiary basis. But I believe it is on this issue of the nature of the consequences of the use of pleasure producing drugs that rational controversy concerning the evaluation of their use must largely turn, and not on philosophical and evaluative differences about the intrinsic value or disvalue of the use of drugs for pleasure.

References and Notes

[1]Gilbert Ryle in *The Concept of Mind*, London, Hutchinson & Co., 1949, criticized extensively the view that pleasure is some kind of sensation or feeling, and few recent philosophers have defended a form of the property of conscious experience theory. J. J. C. Smart in *An Outline of Utilitarian Ethics*, Carlton, Melbourne University Press,. 1961, reprinted in a revised edition in J. J. C. Smart and Bernard Williams, *Utilitarianism: For and Against*, Cambridge, Cambridge University Press, 1973, seems committed to such a view. Other philosophers, such as J. C. B. Gosling, *Pleasure and Desire*, Oxford, Oxford University Press, 1969, and Anthony Kenny, *Action, Emotion and Will*, London, Routledge & Kegan Paul, 1963, argue that there is one sense of pleasure in which it is a sensation, but that not all pleasure is a sensation. Part of this theory's importance is that probably most ordinary persons implicitly hold some version of it, and so it in turn underlies such persons' attitudes towards the use of drugs for pleasure.

[2]Among the more extensive discussions of pleasure sympathetic to some form of the preference theory are J. L. Cowan, *Pleasure and Pain*, New York, St. Martins Press, 1968, and D. L. Perry, *The Concept of Pleasure*, The Hague, Mouton, 1967. An excellent, shorter review of alternative theories of pleasure is William Alston, "Pleasure," in Paul Edwards, ed., *The Encyclopedia of Philosophy*, New York, Macmillan, 1967.

[3]Smart and Williams, p. 19.

[4]John Stuart Mill, *Utilitarianism*, S. Gorovitz, ed., Indianapolis, Bobbs-Merrill, 1971, pp. 18–19.

[5]For readers unfamiliar with the philosophical literature and positions on the mind–body problem, two useful starting points are Richard Taylor, *Metaphysics*, Englewood Cliffs, NJ, Prentice-Hall, 1963, Chapters 1–3, and the entry, "The Mind–Body Problem," by Jerome Schaffer

in Paul Edwards, ed., *The Encyclopedia of Philosophy*, New York, Macmillan, 1967.

⁶Taylor and Schaffer also contain introductory discussions of materialism and physicalism.

⁷For a general discussion and evaluation of ethical hedonism, see Richard Brandt, *Ethical Theory*, Englewood Cliffs, Prentice-Hall, 1959, Chapter 12, 13.

⁸Smart and Williams, p. 25.

⁹Mill, pp. 19–21. A contemporary defense of hedonism that employs the distinction between different quality pleasures is Rem Edwards, *Pleasures and Pains: A Theory of Qualitative Hedonism*, Ithaca, Cornell University Press, 1979.

¹⁰A recent statement and defense of a happiness theory of the good for persons can be found in Richard Brandt, *A Theory of the Good and the Right*, Oxford, Oxford University Press, 1979.

¹¹A recent statement and defense of the desire theory of the good for persons can be found in John Rawls, *A Theory of Justice*, Cambridge, Harvard University Press, 1971, Ch. 7.

¹²See, for example, the criticisms of desire theories made by Brandt, *Good and the Right*, pp. 247–253.

¹³Systematic statements and defenses of perfectionist theories are difficult to find in recent philosophical literature. Nevertheless, many persons are committed to some form of perfectionist theory, as I have construed such theories in the text, by their substantive views about the good for persons. For example, anyone who holds that it is good for a person not to read pornography and not to engage in exploitative personal relations, even if that person desires to or is made happy by doing so, is committed to what I have called a perfectionist theory.

Drugs, Sports, and Ethics

Thomas H. Murray

Our images of the nonmedical drug user normally include the heroin addict nodding in the doorway, the spaced-out marijuana smoker, and maybe, if we know that alcohol is a drug, the wino sprawled on the curb. We probably do not think of the Olympic gold medalist, the professional baseball player who is a shoo-in for the Hall of Fame, or the National Football League lineman. Yet these athletes and hundreds, perhaps thousands, of others regularly use drugs in the course of their training, performance, or both. I am talking not about recreational drug use—athletes use drugs for pleasure and relaxation probably no more or less than their contemporaries with comparable incomes—but about a much less discussed type of drug use: taking drugs to enhance performance.

It is a strange idea. Most of us think of drugs in one of two ways. Either they are being properly used by doctors and patients to make sick people well or at least to stem the ravages of illness and pain, or they are being misused—we say "abused"—by individuals in pursuit of some unworthy pleasures. Performance-enhancing drug use is so common and so tolerated in some forms that we often fail to think of it as "drug" use. The clearest example is the (caffeinated) coffee pot, which is as much a part of the American workplace as typewriters and timeclocks. We drink coffee (and tea and Coke) for the "lift" it gives us. The source of "that Pepsi feeling" and the "life" added by Coke is no mystery—it is caffeine or some of its close chemical relatives, potent stimulants to the human central nervous system. Anyone who has drunk too much coffee and felt caffeine "jitters," or drunk it too late at night and been

107

unable to sleep can testify to its pharmacological potency. Caffeine and its family, the xanthines, can stave off mental fatigue and help maintain alertness, very important properties when we are working around a potentially dangerous machine, fighting through a boring report, or driving for a long stretch. In other words, caffeine can enhance our ability to perform tasks that would otherwise be so fatiguing that we might do them badly, or even harm ourselves or others in trying.

But is caffeine a drug? The fact that it is not under the control of doctors is irrelevant. It was only in 1938 that a significant number of drugs were placed under medical control. The contemporary classification into prescription and nonprescription drugs is very recent.[1]

Does it matter that caffeine is found in natural substances? Certainly not. The first antibiotics were simply purified forms of substances produced naturally by organisms in the soil. Early herbal and dietary remedies, for all their faults, often turn out to contain substances that we have added to the modern pharmaceutical armory. About the only definition of "drugs" that would exclude things like caffeine and alcohol would be one that arbitrarily excludes them because of their wide availability and long and extensive history of use. Any reasonable definition of "drug" based on its effects on the human organism would have to include these two as well as nicotine and a number of other common substances.

Caffeine, then, is a performance-enhancing drug. Using caffeine to keep alert is an instance of the nonmedical use of a drug. So, too, is consuming alcohol at a cocktail party for the pleasure of a mild inebriation, or as a social lubricant to enable you to be charming to people you find intolerably boring when you are sober. In the first case, alcohol is a pleasure-enhancer; in the second, it is a performance-enhancer. What the drug is used for and the intention behind the use—not the substance itself—determines whether we describe it as medical or nonmedical; as pleasure-, performance-, or health-enhancing.

Drugs on the Playing Field

The area of human endeavor that has seen the most explosive growth in performance-enhancing drug use is almost certainly sport. At the highest levels of competitive sports, where athletes strain to improve performances already at the limits of human ability, the temptation to use a drug that might provide an edge can be powerful. Is this kind of drug use unethical? Should we think of it as an expression of liberty? Or do the special circumstances of sport affect our moral analysis? In particular, should liberty give way when other important values are threatened, and when no one's good is advanced? These questions frame the discussion that follows.

Although reports of drug use, perhaps in the form of herbs and mushrooms, date back as far as the Greek Olympiads, the modern applications began in the late nineteenth century, with preparations made from the coca leaf—the source of cocaine and related alkaloids. Vin Mariani, a widely used mixture of coca leaf extract and wine, was even called "the wine for athletes." It was used by French cyclists and, according to W. Golden Mortimer's book *Peru: History of Coca*, by a champion lacrosse team. Coca and cocaine were popular because they staved off the sense of fatigue and hunger brought on by prolonged exertion.[2]

But drugs did not enter sport in a major way until the 1960s. The 1964 Olympics were probably the first in which steroids were used. A group of steroids related to the masculinizing hormone testosterone was synthesized. These steroids are valued for two principal effects, which they have in varying proportions: androgenic, or masculinizing, and anabolic, or tissue-building. Anabolic steroids—those with preponderantly tissue-building impact—were originally used to help repair damaged tissue, including skeletal muscles. Some unknown athlete must have reasoned that if anabolic steroids can rebuild damaged muscles, why not use them to build additional normal muscle?

Among international athletes, the practice became very popular very rapidly. In 1973, Gold Medalist and four-time Olympic competitor in the hammer-throw, Harold Connelly, testified before a US Senate Committee: "It was not unusual in 1968 to see athletes with their own medical kits, practically a doctor's, in which they would have syringes and all their various drugs. . . . I know any number of athletes on the 1968 Olympic team who had so much scar tissue and so many puncture holes in their backsides that it was difficult to find a fresh spot to give them a new shot."[3]

The situation is no better, and probably is much worse now. In the Montreal Olympics in 1976, observers noted the East German women swimmers' deep voices and other vaguely masculine features. Their coach insisted, "We have come here to swim, not sing."[4] When Renate Neufeld, another East German athlete, refused to take "hormone pills" in May 1977, she was ordered to undergo psychotherapy. She later defected to the West.[5] In the summer of 1979, seven athletes from Eastern Europe, including the top three 1500-meter runners, were suspended when steroids were detected in their urine.[6] Nor are all athletes who use steroids from Eastern Europe. A substantial number of Western athletes, including those from the United States, have also been suspended for steroid use.

Those who lift or throw weights appear to have been the first to use steroids, and even today use them most. An informal, confidential survey of weight lifters revealed that between 90 and 100 percent used steroids. A recent study of power-lifters and body builders outside the Olympics, where athletes are a bit less reluctant to talk about their use of performance-enhancing drugs, found almost universal use of steroids as part of training. Olympic hammer thrower, George Frenn, said "I honestly cannot name one guy, and I know just about all of them personally, who is not using steroids." Other reports and confidential sources indicate that steroid use, though less frequent in other events, is common-place among track and field athletes. And it seems to be growing. Harold Connelly recently said, "Athletes now are taking doses that would have blown our minds, and kids are taking them younger."[7]

At about the same time that steroids were introduced into amateur sports, amphetamines and related stimulants were finding their way into professional sports. We know the most about professional baseball and football. But ignorance about other sports is just that—ignorance—and should not be taken to mean that performance-enhancing drugs are not used in those sports. Our knowledge of drug use in football comes largely from the well-publicized activities of psychiatrist Arnold J. Mandell. In several articles and a book, *The Nightmare Season*, Mandell has given his version of the drug-taking behavior of one team in particular—the San Diego Chargers—as well as of professional football in general, and defended his role with the team. Accounts by journalists, supplemented by confidential reports from former professional football players, confirm the general outlines of Mandell's portrait of stimulant use, and even expand it to include other drugs. According to one former player, the first question an injured player is asked before giving him medication is "What else have you taken today?" Mandell reported that defensive linemen, who can play effectively in a state of semicontrolled rage, might take as much as 150 mg of amphetamine in preparation for a game. One player recounted the ritual of another player who would begin taking stimulants many hours before the game, a fixed dose every few hours.[8]

There were drawbacks to the amphetamines. One young player, getting his first start at center, recalls a veteran giving him a couple of amphetamine pills to "get him up for the game." He missnapped the ball three times before the count.[9] A veteran defensive lineman, in a crucial game against the Baltimore Colts, overdid it. On one play, he simply tossed his blocker aside, and was face-to-face with the Colts' quarterback at that time, Johnny Unitas. Unitas bent over, covering the football to prevent a fumble, figuring he was about to receive a violent hit. Confused, the lineman stared at Unitas, looked back at the discarded blocker lying on the ground—and jumped on the blocker. A more pervasive problem with high-dose amphetamine use for a game is the comedown afterwards. Players have described being in an exhausted state for a day or longer, "crashing" hard from the heavy doses of amphetamine.

Amphetamine use in professional baseball is less publicized, although at least one reporter has described a scene from the late sixties in the Washington Senators' locker room where a cereal bowl full of amphetamines was available to the players.[10] And at least one likely Hall of Fame candidate, who is still playing, makes a clear allusion to amphetamines, implying he used them to keep up his concentration during the long season.

Although steroids in amateur athletics and stimulants in the professional world are the best known examples of performance-enhancing drugs in sport, they are not the only instances, and their use may be changing. Football players have reported taking as many as six or seven Empirin 4 before a game. If their reports are accurate, this amounts to 360–420 mg of codeine. They claim it gives them a feeling of euphoria, and dulls the pain from high-speed collisions typical for players on the so-called special teams. We have heard reports that in a nationally televised program in which men competed in feats of strength, they used a series of carefully titrated pharmaceuticals, capped by a whiff of that old reliable, but short-acting stimulant—cocaine. This may be the epitome of athletic drug enhancement, both in quantity and variety. Typically, use is probably limited to one or two drugs, and many athletes, even in the sports most linked to drug use, certainly abstain.

Amphetamine use has decreased as supplies and official pressure to control use has increased. Steroid use continues in its own special kind of competition between the athletes who try to remain one step ahead of the "Doping Control" agents and Olympic authorities. In the 1980 Lake Placid Winter Olympics alone, drug testing cost $1.5 million. More sophisticated tests will make it difficult to continue to use steroids, but there is no guarantee that the athletes will not find novel methods to avoid detection.

Why Athletes Take Anabolic Steroids

Many athletes persist in using performance-enhancing drugs despite official disapproval, possible disqualification, and even risk to their own health. They do so in the face of expert opinion that casts doubt on the effectiveness of the drugs they take. What leads them to jeopardize their futures as athletes, and possibly their very health, for what some medical people claim is an illusory advantage? The best case to illuminate these questions is steroid use among Olympic athletes. Nowhere else are the penalties greater, the efficacy more contended, or the possible health effects more serious.

Before trying to answer the question directly, we need to examine the evidence on the actual effects—desired and undesired—of the anabolic steroids. How are athletes using these drugs, and with what impacts?

The *Physicians Desk Reference* lists five drugs as anabolic hormones. All are testosterone-related steroids; all are widely known within the community of athletes. Most carry the following caution: "Warning: Anabolic steroids do not enhance athletic ability." Recommended therapeutic dosages for the oral forms (Winstrol, Anavar, Dianabol) vary from 2.5 to 10 mg per day. A typical dosage for an athlete in training is 100 mg per day and up, with as much as 400 mg per day claimed. The oral forms are used along with injectables such as Decadurabolin, which may be taken twice a week, rather than once every three or four weeks, as recommended for therapy.

Thus, anabolic steroids are being taken in quantities as much as twenty times greater than the therapeutic dose. This suggests two things. First, the information gained about anabolic effects based on doses typically used in therapy is only marginally relevant to the actual patterns of use among athletes. Claims by medical experts about the lack of efficacy of anabolic steroids in building athletic ability are based on those lower dosages, and are of doubtful validity. Second, our knowledge about side effects, especially chronic ones, is also based on these low dosages; at higher dosages some side effects may be more severe and others may first appear. Under the current federal regulation of human subjects research no institutional review board would approve a research design that entailed giving subjects anywhere near the levels of anabolic steroids used by the athletes. Fortunately, however, research on the effects on chronic high-dose users has begun, on bodybuilders and power lifters.[11]

Though some doctors are doubters, athletes are convinced that anabolic steroids, in conjunction with strenuous training, significantly enhance their performance. Athletes have testified to sudden increases in ability—precipitous drops in sprint times, rapid increases in the length of throws. The achievements have been particularly noticeable in women who have suddenly undergone masculinization. One major critic of steroid use claims that the scientific evidence does not support the hypothesis that steroids lead to performance gains; and that athletes misinterpret a weight increase, which is probably caused by water retention, as a gain in muscle.[12] Although he is correct about the lack of direct scientific proof besides the anecdotal evidence, there is also indirect scientific evidence of their effectiveness.

A recent report in *The Lancet* acknowledged, "Unquestionably, anabolics improve live weight gain, carcass weight, feed efficiency, and percentage meat in some species."[13] Although the report was referring to the practice of mixing steroids with livestock feed, it might have referred equally to humans. Aside from the genitals, the major active site for anabolic–androgenic steroids is skeletal muscles.[14] The action of steroids on muscle tissue is probably related to their effect on nitrogen metabolism—dietary nitrogen is utilized more efficiently in androgenized tissue. There is even suggestive evidence that androgens may act directly on heart tissue; androgen receptors are found in atrial and ventrical myocardial cells.

Athletes also believe that anabolic–androgenic steroids make them more aggressive, and thus enable them to train harder. The case for this, as with most behavioral rather than physiological effects, is more elusive. There is a reasonable amount of evidence that prenatal exposure to the prototypic anabolic androgen—testosterone—can masculinize certain behavior in females. A similar effect appears in a genetic anomaly known as congenital adrenal hyperplasia (CAH). In this condition female fetuses are exposed to abnormally high levels of testosterone. Even though the condition is treatable after birth to reduce overt masculinization, the girls exhibit masculine behavior. [15]

These examples, though, do not establish what impact increased androgenic steroids would have on adult men and women. The evidence here is inconclusive. The effect noticed by the athletes could be as easily explained by the increased efficiency in muscle nitrogen metabolism, which could lessen fatigue in the midst of training regimens.

An Unsightly Array of Risks

Along with the effects athletes desire to obtain from steroids come others not so welcome. Again, the lack of careful scientific information on chronic, high-dose usage forces us to rely on anecdotal reports and reasonable inferences. The anecdotes can be frightening. It is important to bear in mind that, despite being billed as "anabolic" steroids, all these drugs are related to testosterone and have a mix of anabolic and androgenic effects. Tests to estimate the ratio of anabolic to androgenic activity may be useful as very rough guidelines, but the evidence indicates that the effects on individual organ systems may vary widely for any given steroid, and that the distinction between "anabolic" and "androgenic" is greatly oversimplified. [16] Therefore, any potent anabolic is also going to have a mix of androgenic effects. In male athletes this leads to a well-known decrease in fertility, since the synthetic androgen interferes with the normal feedback loop linking endogenous testosterone with sperm production. Other effects include acne (a real give-away for steroid use, according to one athlete), and less visibly, cholesterol buildup and altered liver function. Athletes today are likely to have located a doctor who is willing to help them monitor some of the more subtle effects.

The impact on female athletes, like the East German swimmers, is more visible, and potentially catastrophic. Patricia Connelly, former Olympian and now coach, has lamented the use of steroids by women saying it robs them of their womanliness. The obvious changes—lowered voice, increased body hair, masculinized build—are probably reversible. Other changes are less well understood and potentially more dangerous. I have heard reports of two women, world champions in the 1970s, who appeared to age with stunning rapidity. One—in her early thirties—began to grow bald, developed age spots and wrinkles on her skin. The second, who also looked very old, lost virtually all her hair and wore a wig. Although no conclusive proof is possible that these effects were caused by steroids, their women competitors have no doubt that steroids were the cause.

Facing such an unpleasant, poorly understood, and unsightly array of risks, why do athletes persist in using steroids? The answer lies in the nature of international, or for that matter professional, athletic competition. In his Senate testimony, Harold Connelly said " . . . the overwhelming majority of the international track and field athletes I have known would take anything and do anything short of killing themselves to improve their athletic performance." The pressures are almost as intense in professional football. As one former player describes it: "It's hard to get violent every Sunday at 1 PM." Amphetamines permitted athletes to play with the necessary aggressive intensity. The same player said, "Every team was looking for an edge." Another footballer who described himself as the forty-fifth man on a forty-five man roster explained: "If you were a star, you could let your injuries heal naturally, But hey, I've got to take the shot. I didn't want the coach to wonder why I hadn't played the last three games and say, 'Hey, he's 33 years old. Let's get a new player.' That's why marginal players do it. We know we're hurting our bodies."[17] Other athletes confirm the tone and substance of these remarks. The intense competition at the highest levels of sport calls for every effort the athletes can make, and pushes them to seek every possible advantage over their competitors. Olympic shot-putter, Al Feuerbach may have put it best in his ironic doggerel on Dianabol:

> Dianabol, Dianabol,
> It's the gateway to fame.
> With Dianabol you'll win them all
> Unless the others are using the same.

In a competitive endeavor, participants will be pressed to use any means available to achieve a competitive advantage. The higher the stakes, the more intense the competition, the more total the commitment, the more likely people are to use exceptional means. For the international amateur or professional athlete this has often come to include drugs. The pressure to use exists when people *believe* that something confers a competitive advantage, whether or not this is objectively true. There is, then, an *inherent coerciveness* present in these situations: when some choose to do what gives them a competitive edge, others will be pressed to do likewise, or resign themselves to either accepting a competitive disadvantage, or leaving the endeavor entirely.

Although sport is not an adequate metaphor for life in most respects, the phenomenon of inherent coerciveness is almost certainly not limited to sport. It is a general feature of society where competition dominates a number of settings, especially work. (Can there be any doubt that for the athletes I have described, their sport is preeminently a form of work?) Where work is highly competitive, whatever confers an advantage, or seems to, will be used by some, and others will thereby be pressed to use those means themselves, or accept the equivalent of a five-yard handicap in the hundred yard dash.

In work, apart from playing fields and gymnasiums, performance-enhancing drugs, especially stimulants, are commonly used. Truck drivers who must remain alert despite the monotony of hours and miles have long used a variety of substances for help. College and professional-school students have used their share of No-Doz and more powerful drugs. And of course the ubiquitous coffee pot in workplaces helps many of us wake up in the morning and stay awake through the afternoon doldrums. These seem innocuous enough.

But the relation of drugs to work has not always been benign. The early history of coca-leaf chewing in South America is informative. The Incas reserved coca-chewing for their notables and for the masses on special religious occasions or as medicine. At first the Spanish disapproved of coca, probably for its link with the Inca religion. But after mine and plantation owners discovered that Indian laborers who chewed coca would work longer and harder without complaining, they persuaded the clerical authorities to lift their opposition. Even today coca use is heaviest among the illiterate, desperately poor manual laborers in the Andean valley, who associate it with work. When a physician, Dr. Carlos Gutierrez-Noriega, criticized the economic interests who encouraged coca use among their Indian workers, he was forced to leave his post at the University of Lima.[18] Not for a moment do I believe that the position of a Peruvian silver miner is equivalent to that of an Olympic or professional athlete in the US. But the silver miner's plight should alert us to the potential dangers when work becomes so constructed that people find it necessary to use drugs to function effectively.

An Ethical Account

Now we can confront the fundamental ethical question: May athletes use drugs to enhance their athletic performance? The International Olympic Committee has given an answer of sorts by flatly prohibiting "doping" of any kind. This stance creates at least as many problems as it solves. It requires an expensive and cumbersome detection and enforcement apparatus, turning athletes and officials into mutually suspicious adversaries. It leads Olympic sports medicine authorities to proclaim that drugs like steroids are ineffective, a charge widely discounted by athletes, and thereby decreases the credibility of Olympic officials. Drug use is driven underground, making it difficult to obtain sound medical data on drug side-effects.

The enforcement body, in an attempt to balance firmness with fairness, bans athletes "for life" only to reinstate them a year later, knowing that what distinguishes these athletes from most others is only that they were caught. A blanket prohibition fails to distinguish legitimate *therapeutic* drug use from use for performance-enhancement. Rick DeMont, an asthmatic swimmer whose Olympic medal was denied when it was discovered that his asthma inhaler contained a banned substance, may have been a victim of this last confusion. Would we deny insulin to a diabetic or aspirin to someone with a severe headache on the grounds that they are "doping?"

In short, the IOC's position has probably had little impact on steroid use. It may make the IOC look good to the public, but it has driven a broad wedge between itself and the athletes it presumably represents. In any case, the IOC does not bother to give reasons for its drug policy, presumably because it assumes the reasons are self-evident, which they are not.

Any argument for prohibiting or restricting drug use by Olympic athletes must contend with a very powerful defense of such use based on our concept of individual liberty. We have a strong legal and moral tradition of individual liberty that proclaims the right to pursue our life plans in our own way, to take risks if we so desire and, within very broad limits, to do with our own bodies what we wish. This right in law has been extended unambiguously to competent persons who wish to refuse even life-saving medical care. More recently, it has been extended to marginally competent persons who refuse psychiatric treatment. Surely, competent and well-informed athletes have a right to use whatever means they desire to enhance their performance.

Those who see performance-enhancing drug use as the exercise of individual liberty are unmoved by the prospect of some harm. They believe it should be up to the individual, who is assumed to be a rational, autonomous, and uncoerced agent, to weigh probable harms against benefits, and choose accordingly to his or her own value preferences. It would be a much greater wrong, they would say, to deny people the right to make their own choices. Why should we worry so much about some probabilistic future harm for athletes when many other endeavors pose even greater dangers? High-steel construction work and coal-mining, mountain-climbing, hang-gliding, and auto-racing are almost certainly more dangerous than using steroids or other common performance-enhancing drugs.

Reasons commonly given to limit liberty fall into three classes: those that claim that the practice interferes with capacities for rational choice; those that emphasize harms to self; and those that emphasize harms to others. The case of performance-enhancing drugs and sport illustrates a fourth reason that may justify some interference with liberty, that given the social nature of the enterprise, performance-enhancing drug use in sport is inherently coercive. But first the other three reasons.

There is something paradoxical about our autonomy: we might freely choose to do something that would compromise our future capacity to choose freely. Selling yourself into slavery would be one way to limit liberty, by making one's body the property of another person. If surrendering autonomous control over one's body is an evil and something we refuse to permit, how much worse is it to destroy one's capacity to *think* clearly and independently? Yet that is one thing that may happen to people who abuse certain drugs. We may interfere with someone's desire to do a particular autonomous act if that act is likely to cause a general loss of the capacity to act autonomously. In this sense, forbidding selling yourself into slavery and forbidding the abuse of drugs likely to damage your ability to reason are similar *restrictions* on liberty designed to *preserve* liberty.

This argument applies only to things that do in fact damage our capacity to reason and make autonomous decisions. Although some of the more powerful pleasure-enhancing drugs might qualify, no one claims that performance-enhancing drugs like the steroids have any deleterious impact on reason. This argument, then, is irrelevant to the case of performance-enhancing drugs.

Destroying one's reasoning ability is a special kind of harm to oneself, but there are many other kinds of harm, and they constitute a potential argument in favor of curtailing liberty. Aristotle, in the Nichomachean Ethics, described a conception of *eudaimonia,* or the good life, in which the perfection of natural excellences was a central component. Our physical abilities, character traits, and above all our intellect were all to be perfected. We can infer from this that persons have a duty to do whatever is in their power to perfect their talents, and forsake whatever would interfere with that development. Aristotle might have no hesitation in condemning most or all pleasure-enhancing drug use, but his principle, which paraphrased is "People should develop their natural excellences," stumbles over the case of athletes and performance-enhancing drugs. Athletes use the drugs precisely to perfect their natural excellences. Our objection would have to be that drug use is an improper means to that end, since the end itself is especially commendable in the Aristotelian worldview. We could argue that drug use is wrong because, on balance, it hurts the pursuit of excellence more than it helps it. But then we are reduced to arguing about the facts—improved performances versus side effects. And if it turns out that the drugs work with minimal, reversible side effects, we would say they are morally justified under the perfection-of-excellence principle.

Another way to put the Aristotelian objection is that we should not use unnatural means—drugs—to perfect natural excellences—athletic abilities. It is difficult to make this stick as a *moral*, and not merely an esthetic, objection. Although we recognize that certain means of perfecting our natural excellences are generally regarded as illegitimate and may be dismissed as "unnatural," the judgments are always with respect to specific ends. A prosthetic hand might be "unnatural" and unfair for the purpose of pitching a baseball, but unobjectionable if it allowed an injured novelist to operate a typewriter. Even in so narrow a case as drugs in sport, all depends on context. If we would not object to a diabetic athlete using insulin, could we object to a depressed one using an antidepressant, to an exhausted one using a stimulant? We can draw lines, but based on complex practical understandings, rather than solid, simple principles delineating the "natural" from the "unnatural."

Allegations of unnaturalness are complicated and difficult to sustain, and not the central evaluative category. The important question is whether or not in specific circumstances the use of a particular technique, substance, and so on is appropriate for the purpose in question. I suspect most people would agree that the use of a chemical stimulant (though possibly not an antidepressant and certainly not insulin) is inappropriate in athletic competition for the purpose of improving one's performance. That same stimulant might be permissible in other circumstances (driving a car) for other purposes (avoiding accident or injury from fatigue). We care whether it is appropriate and justifiable, not whether it is natural or unnatural. In other words, there is nothing intrinsically immoral about using "unnatural" means to an end; everything will depend on questions of context, consensus, and purpose.

Aristotle's concept of eudaimonia, with its emphases on intellectual *and* physical excellence, seems particularly appropriate for considering the problem of performance-enhancing drugs and sport, although it is certainly not the only moral theory that suggests that we may have moral obligations to ourselves. But any theory of moral obligations to self would have to admit that taking risks to one's health is inescapable, and that the question becomes one of proportionality: whether the risk is proportionate to the goal. Athletes taking heavy doses of steroids are accepting considerable risks. But they, and perhaps the hundreds of millions of people who follow the Olympics, regard the goal as anything but trivial.

A second class of reasons to limit liberty says that we may interfere with some actions when they result in wrong to others. The wrong done may be direct—lying, cheating, or other forms of deception are unavoidable when steroid use is banned, yet one persists in the practice. Of course, we could lift the ban, and then the steroid use need no longer be deceptive; it could be completely open to the same extent as other training aids. Even with the ban forcing steroids into the pale of secrecy, it would be naive to think that other athletes are being deceived when they all know that steroids are in regular use. The public may be deceived, but not one's competitors. It is hard to lie when no one is deceived. Using steroids may be more like bluffing in poker than fixing the deck, at least for your competitors.

The wrong we do to others may be indirect. We could make ourselves incapable of fulfilling some duty we have to another person. For example, a male athlete who marries and promises his wife that they will have children makes himself sterile with synthetic anabolic steroids (a probable side effect). He has violated his moral duty to keep a promise. This objection could work, but only where the duty is clearly identifiable and not overly general, and the harm is reasonably forseeable. Duties to others must have a limited, clear scope or become absurdly general or amorphous. We may believe that parents have the duty to care for their children, but we do not require parents to stay by their children's bedside all night, every night, to prevent them from suffocating in their blankets. The duty is narrower than that. In order to avoid being too vague, we would have to be able to specify what the "duty to care for one's children" actually includes. Except for cases like the sterile athlete reneging on his promise, instances where athletes make themselves incapable of fulfilling some specific duty to others would probably be rare. In any case, we cannot get a general moral prohibition on drug use in sports from this principle, only judgments in particular cases.

We may also do a moral wrong to others by taking unnecessary risks and becoming a great burden to family, society, or both. The helmetless motorcyclist who suffers severe brain damage in an accident is a prototypical case. Increasingly, people are describing professional boxing in very similar terms. Although this might be a good reason to require motorcyclists to wear helmets or to prohibit professional boxing matches, it is not a sound reason to prohibit steroid use. No one claims that the athletes using steroids are going to harm themselves so grievously that they will end up seriously brain-damaged or otherwise unable to care for themselves. Though the harms they do to themselves may be substantial, they are not disabling.

"Free Choice Under Pressure"

So far we have not found any wrong done to others that is serious and likely to occur when athletes, at the top of the competitive ladder, commonly use performance-enhancing drugs. Let us look at the problem more closely. Olympic and professional sport, as a social institution, is an intensely competitive endeavor, and there is tremendous pressure to seek a competitive advantage. If some athletes are believed to have found something that gives them an edge, other athletes will feel pressed to do the same, or leave the competition. Unquestionably, coerciveness operates in the case of performance-enhancing drugs and sport. Where improved performance can be measured in fractions of inches, pounds, or seconds, and that fraction is the difference between winning and losing, it is very difficult for athletes to forego using something that they believe improves their competitor's performance. Many athletes do refuse; but many others succumb; and still others undoubtedly leave rather than take drugs or accept a competitive handicap.

Under pressure, decisions to take performance-enhancing drugs are anything but purely "individual" choices. My alleged liberty to take performance-enhancing drugs, which is very hard to oppose from an individualistic conception of morality, is counterbalanced by the pressure I place on my fellow competitors. My "free" choice contains an element of coercion. If enough people like me choose to use performance-enhancing drugs, then the freedom of others not to use them is greatly diminished.

But can we say that "freedom" has actually been diminished because others are using performance-enhancing drugs? I still have a choice whether to participate in the sport at all. In what sense is my freedom impaired by what the other athletes may be doing? If we take freedom or liberty in the very narrow sense of noninterference with my actions, then my freedom has not been violated, because no one is prohibiting me from doing what I want, whether that be throwing the discus, taking steroids, or selling real estate. But if we take freedom to be one of a number of values, whose purpose is to support the efforts of persons to pursue reasonable life plans without being forced into unconscionable choices by the actions of others, then the coerciveness inherent when many athletes use performance-enhancing drugs and compel others to use the same drugs, accept a competitive handicap or leave the competition can be seen as a genuine threat to one's life plan. When a young person has devoted years to reach the highest levels in an event, only to find that to compete successfully he or she must take potentially grave risks to health, we have, I think, as serious a threat to human flourishing as many restrictions on liberty.

At this point it might be useful to consider the social value we place on improved performance in sport. It is a truism that you win a sports event by performing better than any of your competitors. The rules of sport are designed to eliminate all influences on the outcome except those considered legitimate. Natural ability, dedication, cleverness are fine; using an underweight shot-put, taking a ten-yard head start, fielding twelve football players are not. The rules of sport are human-made conventions. No natural law deems that shot-puts shall weigh sixteen pounds, or that football teams shall consist of eleven players. Within these arbitrary conventions, the rules limit the variations among competitors to a small set of desired factors. A willingness to take health risks by consuming large quantities of steroids is *not* one of the desired, legitimate differences among competitors.

Changing the rules of a sport will alter performances, but not necessarily the standing of competitors. If we use a twelve-pound shot-put, everyone will throw it farther than the sixteen-pound one, but success will still depend on strength and technique, and the best at sixteen pounds will probably still be best at twelve pounds. Giving all shot-putters 100 mg of Dianabol a day will have a similar impact, complicated by variations in physiological response to the drug. Noncatastrophic changes in the rules may shift some rankings, but will generally preserve relations among competitors. Changes that do not alter the nature of a sport, but greatly increase the risks to competitors are unconscionable. Changes that affirmatively tempt athletes to take the maximum health risk are the worst. Lifting the ban on performance-enhancing drugs would encourage just that sort of brinksmanship. On the other hand, an effective policy for eliminating performance-enhancing drug use would harm no one, except those who profit from it.

My conclusions are complex. First, the athletes who are taking performance-enhancing drugs that have significant health risks are engaging in a morally questionable practice. They have turned a sport into a sophisticated game of "chicken." Most likely, each athlete feels pressed by others to take drugs, and does not feel he or she is making a free choice. The "drug race" is analogous to the arms race.

Second, since the problem is systemic, the solution must be too. The IOC has concentrated on individual athletes, and even then it has been inconsistent. This is the wrong place to look. Athletes do not use drugs because they like them, but because they feel compelled to. Rather than merely punishing those caught in the social trap, why not focus on the system? A good enforcement mechanism should be both ethical and efficient. To be ethical, punishment should come in proportion to culpability and should fall on *all* the guilty parties—not merely the athletes. Coaches, national federations, and political bodies that encourage, or fail to strenuously discourage, drug use, are all guilty. Current policy punishes only the athlete.

To be efficient, sanctions should be applied against those parties who can most effectively control drug use. Ultimately, it is the athlete who takes the pill or injection, so he or she ought to be one target of sanctions. But coaches are in an extraordinarily influential position to persuade athletes to take or not to take drugs. Sanctions on coaches whose athletes are caught using drugs could be very effective. Coaches, not wanting to be eliminated from future competitions, might refuse to take on athletes who use performance-enhancing drugs.

Finally, although I am not in a position to elaborate a detailed plan to curtail performance-enhancing drug use in sports, I have tried to establish several points. Despite the claims of individual autonomy, the use of performance-enhancing drugs is ethically undesirable because it is coercive, has significant potential for harm, and advances no social value. Furthermore, any plan for eliminating its use should be just and efficient, in contrast to current policies.

Can we apply this analysis of drug use in sports to other areas of life? One key variable seems to be the social value that the drug use promotes, weighed against the risks it imposes. If we had a drug that steadied a surgeon's hand and improved his or her concentration so that surgical errors were reduced at little or no personal risk, I would not fault its use. If, on the other hand, the drug merely allowed the surgeon to operate more quickly and spend more time on the golf course with no change in surgical risk, its use would be at best a matter of moral indifference. Health, in the first case, is an important social value, one worth spending money and effort to obtain. A marginal addition to leisure time does not carry anywhere near the same moral weight.

A careful, case-by-case, practice-by-practice weighing of social value gained against immediate and long-term risks appears to be the ethically responsible way to proceed in deciding on the merits of performance-enhancing drugs.

Acknowledgments

Whatever philosophical value this paper has is largely due to the generous help given by three colleagues: Gregory Vlastos, Margaret Pabst Battin, and Arthur Caplan. I owe them much.

References

[1] Peter Temin, *Taking Your Medicine: Drug Regulation in the United States,* Cambridge, Harvard University Press, 1980.

[2] Lester Grinspoon and James B. Bakalar, *Cocaine: A Drug and its Social Evolution,* New York, Basic Books, 1976.

[3] Barry Lorge, "A Thoroughly Modern Athlete: Bigger, Better—and on Drugs," *Washington Post,* May 27, 1979, p. A6.

⁴Neil Amdur, "Mounting Drug Use Afflicts World Sports," *New York Times,* November 20, 1978, pp. C1, C8.

⁵M. Getler, "Athlete Who Fled E. Germany Cites Forced Drug Use," *Washington Post,* December 29, 1978, pp. A1, A23.

⁶Neil Amdur, "Seven Women Athletes Banned for Drugs," *New York Times,* October 26, 1979, pp. A1, A25.

⁷Lorge, "A Thoroughly Modern Athlete."

⁸Arnold Mandell, *The Nightmare Season,* New York, Random House, 1976.

⁹Gary Smith, "Ex-NFLers Testify Drug Problem is Bigger Than Rozelle Ever Imagined," *New York Daily News,* November 29, 1980.

¹⁰T. Boswell, "Number of 'Poppers' in Baseball Grows Fewer Each Season," *Washington Post,* May 28, 1979.

¹¹R. H. Strauss, J. E. Wright, G. A. M. Finerman, and D. H. Caitlin, "Side Effects of Anabolic Steroid Hormones in Weight-Trained Male Athletes," *Physician and Sports Medicine,* December 1983, pp. 86–95.

¹²A. J. Ryan, "Anabolic Steroids Are Fool's Gold," *Proceedings of the Federation of American Societies for Experimental Biology,* 40 1981, pp. 2682–2688.

¹³Editorial: "Anabolics in Meat Production," *Lancet,* March 27, 1982, pp. 721–722.

¹⁴C. W. Bardin and J. F. Catterall, "Testosterone: A Major Determinant of Sexual Dimorphism," *Science* 211, 1981, pp. 1285–1294.

¹⁵A. A. Ehrhardt, and H. F. L. Meyer-Bahlburg, "Effects of Prenatal Sex Hormones on Gender-Related Behavior," *Science* 211, 1981, pp. 1312–1318.

¹⁶Bardin and Catterall, "Testosterone."

¹⁷Smith, "Ex NFLers Testify Drug Problem is Bigger Than Rozelle Ever Imagined."

¹⁸Grinspoon and Bakalar, *Cocaine.*

¹⁹Aristotle, *The Ethics of Aristotle,* New York, Penguin, 1955.

Part III

Privacy,
the Constitution,
and Drug Use

Implications of the Constitutional Right of Privacy for the Control of Drugs

An Introduction

Robert Schwartz

Introduction

Probably no Constitutional issue has created as much interest in academic circles over the past several years as has the development of the Constitutional right of privacy and, with the exception of the debate over affirmative action, no issue has led to such a volume of both judicial and scholarly writing. There are now several excellent descriptions, analyses, and histories of the right of privacy.[1] The purpose of this survey presentation is to evaluate the development of the Constitutional right of privacy and those social policies that must underlie it in order to determine how the United States Supreme Court might apply the right of privacy to drug prohibitions and regulations and what implications the underlying social policies ought to have for our drug laws.

Although the United States Supreme Court has indicated that it will give very wide berth to state and federal regulation and prohibition of the use of drugs, there is no definitive Supreme Court opinion applying the right of privacy to any kind of drug regulation or prohibition. Thus, it may be helpful to consider the Supreme Court discussion of the right of privacy

in related areas, and the lower courts' application of the right of
privacy in cases challenging drug regulation and prohibition as
well as in analogous cases. In particular, it may be useful to an-
alyze judicial approaches to decisions to undergo or forego particular medical care; attempts to sell, possess, or "use" pornography; state requirements that motorcyclists wear helmets or
automobile drivers wear seatbelts; access to contraception and
abortion; legislatively disfavored sexual practices; unusual
dress; and purely commercial speech. Although no particular
decision in any of these areas will certainly and accurately pre-
dict the law the United States Supreme Court will apply in
resolving questions surrounding drug regulation, decisions in
analogous areas ought to shed light on the role the right of privacy is likely to play in the judicial determination of public drug
policy.

This paper will describe judicial application of the right of
privacy in these areas by briefly reviewing the development of
the right, surveying those two principles—control of mind and
control of body—that must underlie that right, and evaluating
the state interests which might impose limitations on the right.

A Brief History of the Constitutional Right of Privacy

Histories of the right of privacy normally begin with the
proposal of such a right by Warren and Brandeis in the *Harvard
Law Review*.[2] As Warren and Brandeis argue, the right is no
more than a combination of several interests that have always
been recognized at the common law.[3] As early as 1891, in *Union Pacific Railroad Co. v. Botsford*, the Supreme Court announced "[no] right is held more sacred, or is more carefully
guarded, by the common law, than the right of every individual to the possession and control of his own person, free from
all restraint or interference of others, unless by clear and unquestionable authority of law. As well said by Judge Cooley,
'The right to one's person may be said to be a right of complete
immunity: to be let alone.' "[4] Combining the several elements
of the right of privacy and applying a consistent analysis to
them was the genius of Warren and Brandeis, and calling the

combination the right of privacy was clever enough to satisfy another of the authors' interests by attracting attention to the newborn *Harvard Law Review*.

Although their original description of this "right of privacy" left it a creature of tort law, and not compelled by the United States Constitution, Warren and Brandeis were heirs to the 19th Century notions of individualism, personal autonomy, and freedom from traditional constraints that also gave rise to the Fourteenth Amendment. Indeed, the Transcendentalist movement of the 19th Century, the philosophy of Emerson and Thoreau, was founded on a respect for the freedom and autonomy of each individual. It is not surprising that the Transcendentalists were at the forefront of the Abolitionist movement, and it is that prevailing philosophy that can be traced through the post-Civil War amendments outlawing slavery and providing that no "state [shall] deprive any person of life, liberty, or property without due process of law."[5] Thus although the original description of this "right of privacy" left it a creature of tort law, the impulse behind the syntheses of the right clearly sprang from principles then only recently imbedded in the Constitution.

Although the Supreme Court did not formally recognize a right of privacy in the United States Constitution until 1965, some scholars have found the seeds of that right in Justice McReynold's opinion in *Meyer v. Nebraska:* "Without doubt [the 'liberty' of the Fourteenth Amendment] denotes not merely freedom from bodily restraint but also the right of the individual to contract, to engage in any of the common occupations of life, to acquire useful knowledge, to marry, establish a home and bring up children, to worship God according to the dictates of his own conscience, and *generally to enjoy those privileges long recognized at common law as essential to the orderly pursuit of happiness by free men.*"[6] In *Meyer* the Court found that a state statute prohibiting the teaching of any foreign language to children violated the due process clause of the Fourteenth Amendment, even though its purpose was a respectable one, "to foster a homogeneous people with American ideals."[7] Two years later in *Pierce v. Society of Sisters,* the Court applied a similarly broad reading of the due process clause in overturning a state law requiring all children to attend public schools.[8]

Although these two decisions could have rested entirely on some Constitutional notion of privacy, both depended as well on the recognition of the economic interest of the teachers who literally would be put out of business by the state statutes.[9] This doctrine of economic substantive due process, which was often applied to overturn state statutes designed to provide for the improvement of the economic condition or conditions of the work place of employees, read into the Constitution an economic theory unacceptable to most of the country even at the beginning of this century. Indeed, because the Fourteenth Amendment and its due process clause protected such "fundamental liberties" as that to work an excessive number of hours[10] at an especially low salary[11] without the restrictions of "forced" unionization,[12] it is not surprising that the Fourteenth Amendment and its due process clause were looked upon as nothing more than the embodiment of the transient political and economic philosophy of the ruling class.[13]

Given the long and excruciatingly painful terminal illness of economic substantive due process, and its overdue death in the mid-1930s, it is not surprising that the Supreme Court searched for some other foundation for a Constitutional right of privacy. That foundation was revealed in *Griswold v. Connecticut* in which the Court found a Connecticut statute making the provision or use of any contraceptive device a crime to be a violation of the Constitution.[14] Specifically rejecting the Fourteenth Amendment's due process clause as a basis for the decision, Justice Douglas, writing for the Court, depended on the First, Third, Fourth, Fifth, and Ninth Amendments.[15] As the Court explained, "[S]pecific guarantees in the Bill of Rights have penumbras, formed by emanations from those guarantees that help give them life and substance. Various guarantees create zones of privacy."[16] Although the Court did not provide any formal way to discover those penumbras, emanations, and the specific rights they protect, it is clear that they were implicated by the Connecticut statute. In order to enforce the statute, the police would be required to "search the sacred precincts of marital bedrooms for telltale signs of the use of contraceptives," which would be "repulsive to the notions of privacy surrounding the marriage relationship."[17] The Court explained that "[w]e deal with a right of privacy older than the Bill of Rights—older than our political parties, older than our school system."[18] Three concurring justices based their opinion on the Ninth Amendment's provision that "[t]he enumeration in the Constitution, of certain rights, shall not be construed to deny or disparage others retained by the people."[19]

Only Justice Harlan could find the right of privacy in the due process clause of the Fourteenth Amendment.[20] He would have held that the Connecticut statute "violates basic values 'implicit in the concept of ordered liberty.' "[21] Justice Harlan depended on his dissenting opinion in *Poe v. Ullman*, in which the Supreme Court, four years before *Griswold*, had declined to review the same statute on technical procedural grounds.[22] In his dissent in that case, Justice Harlan had announced that

> [D]ue process has not been reduced to any formula; its content cannot be determined by reference to any code. The best that can be said is that through the course of this Court's decisions it has represented the balance which our Nation, built upon postulates of respect for the liberty of the individual, has struck between that liberty and the demands of organized society. If the supplying of content to this Constitutional concept has of necessity been a rational process, it certainly has not been one where judges have felt free to roam where unguided speculation might take them. The balance of which I speak is the balance struck by this country, having regard for what history teaches are the traditions from which it developed as well as the traditions from which it broke. That tradition is a living thing. A decision of this Court which radically departs from it could not long survive, while a decision which builds on what has survived is likely to be sound. No formula can serve as a substitute, in this area, for judgment and restraint.[23]

Dissenting in *Griswold*, Justice Black announced that "I like my privacy as well as the next one, but I am nevertheless compelled to admit that government has a right to invade it unless prohibited by some specific constitutional provision."[24] Thus, it was not the fundamental nature of any particular right that was the subject of the disagreement with the majority; rather, it was the Court's arrogation of the determination of what constituted a fundamental right. Justice Black simply did not believe that was part of the judicial function.

Within a decade, the Court suggested that the family relationship was not the dispositive element in *Griswold*, when, in *Eisenstadt v. Baird*, the Court extended the right to use contraceptives to unmarried adults.[25] Justice Brennan announced that "[i]f the right of privacy means anything, it is the right of the *individual*, married or single, to be free from unwarranted governmental intrusion into matters so fundamentally affecting a person as the decision whether to bear or beget a child."[26] Perhaps, then, the more significant element of privacy in *Griswold* was the location of the act—the private chambers of the home. Such a conclusion would be supported by *Stanley v. Georgia*, in which the Court found that there was a right to possess, in one's own home and for one's own use, an obscene movie that would not be the subject of First Amendment protection in the stream of commerce, or outside of the home.[27]

It was not until the abortion cases in 1973 that the United States Supreme Court was able to overcome memories of economic substantive due process and find the right of privacy in the due process clause of the Fourteenth Amendment.[28] Justice Blackmun, speaking for the Court in *Roe v. Wade*, reviewed the history of the Constitutional right of privacy and concluded that "[t]hese decisions make it clear that only personal rights that can be deemed 'fundamental' or 'implicit in the concept of ordered liberty' . . . are included in this guarantee of personal privacy."[29] The Court was willing to define the right of privacy in such a way that it included a woman's right to an abortion, at least during some stages of the pregnancy, even though the abortion itself was not an "intimate" activity, even though it did not take place in the home, and even though the Court recognized that

> When those trained in the respective disciplines of medicine, philosophy, and theology are unable to arrive at any consensus, the judiciary, at this point in the development of man's knowledge, is not in a position to speculate as to the answer.[30]

Although the Court did not attempt any precise taxonomy of privacy in the abortion cases, Justice Douglas, concurring in *Doe v. Bolton*, outlined three classes of "time-honored rights, amenities, privileges, and immunities that come within the sweep of 'the blessings of liberty' mentioned in the Preamble of the Constitution" and thus within the scope of the Fourteenth Amendment:

> First is the autonomous control over the development and expression of one's intellect, interests, tastes, and personality.[31]
> Second is freedom of choice in the basic decisions of one's life respecting marriage, divorce, procreation, contraception, and the education and upbringing of children.[32]
> Third is the freedom to care for one's health and person, freedom from bodily restraint or compulsion, freedom to walk, stroll, or loaf.[33]

The decision in the abortion cases launched a campaign in Congress and the state legislatures to enact legislation that would prohibit or discourage abortions, and these, in turn, have set off an avalanche of Constitutional litigation to determine the limits of the right of privacy.[34] Several cases have considered the form in which the right to an abortion must be extended to children,[35] and the obligations of government agencies to provide or pay for abortions specifically permitted in *Roe v. Wade*.[36]

The United States Supreme Court has twice dealt with privacy challenges to drug regulation, and those cases deserve mention. In 1977, in *Whalen v. Roe*, the Supreme Court upheld a New York statute that required that the name and address of any person prescribed certain federally regulated drugs be maintained in a central file kept by the state.[37] The statute put severe restrictions on access to the information.[38] The plaintiffs, both physicians and patients, argued, in part, that the New York statute would encourage some patients not to accept prescriptions for the regulated drugs that were the subject of the law.[39] This, they argued, would be a severe disruption of the doctor-patient relationship and the privacy implicit in that relationship that was recognized in *Roe v. Wade*.[40] The Court unanimously rejected the challenge, suggesting, in passing, that the state could have prohibited access to those same drugs altogether.[41] Although the constitutional propriety of a complete ban was not before the Court in this case, the gratuitous announcement, and the reputation of the abuse of the regulated drugs (amphetamines and opium derivatives) made it clear that the Court would be deferential to prohibition and regulation of potentially dangerous drugs.

In *Rutherford v. United States*, the Court was called upon to adjudicate a formal right of privacy challenge to the Food, Drug, and Cosmetic Act safety and efficacy standard in a legal challenge to the agency's prohibition on the interstate movement and importation of laetrile.[42] The District Court had specifically based its determination that laetrile could not be effectively prohibited on a constitutionally based right of privacy.[43] The Court of Appeals, reviewing the District Court, affirmed, but on other grounds. The Court of Appeals ruled that the safety and efficacy requirements of the Food, Drug, and Cosmetics Act had no meaning when applied to certified terminally ill patients for whom no medical care could be effective.[44] The United States Supreme Court reversed on the grounds that

the Court of Appeals misread the administrative scope of the
Act, and the Supreme Court, which had heard formal argument on the privacy issues, remanded the case so that the
Court of Appeals could reconsider it on privacy grounds.[45] On
remand, the Court of Appeals refused to recognize the right of
privacy as extending to the use of laetrile.[46] The Supreme
Court, finally faced with a challenge to drug regulation based
on the right of privacy, denied certiorari and refused to hear
the case.[47] The Supreme Court's decision is technically without precedential value, but it demonstrates the Court's reluctance to dispute the nature of drug regulation with the legislative or administrative regulators. Months before the Supreme
Court denied certiorari in *Rutherford*, the Court also denied certiorari in *Privitera v. California*.[48] In that case a California physician had been convicted of conspiracy, a felony, for being part
of a scheme that resulted in the provision of laetrile to several
terminally ill cancer patients. The California Court of Appeal
had reversed his conviction on the basis of both the patient's
and the physician's right of privacy.[49] In a lengthy and moving
opinion, the Court of Appeal declared that

> *[no] demonstrated public danger, no compelling interest of
> the state, warrants an Orwellian intrusion into the most private
> zone of privacy . . .*
> *No compelling interest of the state requires Dr. Privitera's
> nineteen cancer patients to endure the unendurable, to die, even
> forbidden hope.*[50]

The California Supreme Court reversed the Court of Appeal
and reinstated Dr. Privitera's felony conviction.[51] Chief Justice
Bird, herself a cancer victim, dissented and adopted as her
opinion the Court of Appeal opinion. Although the Supreme
Court's denial of certiorari was not a substantive decision on
the merits of the case, it again illustrates the Court's hesitation
to intervene in the state regulation of drugs.

Principles Underlying the Constitutional Right of Privacy

Whatever is encompassed by the right of privacy, that
right includes whatever is "fundamental to our concept of ordered liberty." This constantly repeated description, along

with the logic of the Constitution, dictate that the Constitution must protect any right necessary for the enjoyment of those rights more specifically enumerated. Some rights must be so fundamental that they are the premises necessary to the existence of a society that would adopt our Constitution. This essentially natural law theory, although the subject of some recent legal scholarship,[52] must have been far more congenial to the framers of the Constitution (and to the drafters of the Fourteenth Amendment after the Civil War) than it would be to the 20th Century community.

Whatever else may be necessary for the enjoyment of Constitutional rights, it is clear that our Constitution protects *persons*. In the absence of a notion of personhood, none of the rest of the Constitution has any meaning. Thus, we must determine what attributes of personhood are essential for other rights protected in the Constitution to be significant. There are, it seems, two such fundamental attributes. The first is the control of one's mind, and the second is the control of one's body.

Control of the Mind

There can be no doubt that most of the Bill of Rights has meaning only to sentient beings. Free speech is not protected by the First Amendment so that every community will be guaranteed a share of human-produced noise. It is guaranteed so that ideas can be communicated, so that we will be able to provide our ideas to others, and so that we can digest the ideas provided by others. Similarly, the protection of the press does not have as its purpose the dissemination of meaningless written signs, nor is the right to assemble simply the right to be in the same general location with other people. All of these rights depend on our interest in having lucid, intentional expressions made in such a way that they affect the way others think. It would make no sense for the Constitution to absolutely protect the input and output of the mind, but not protect the mind's internal function at all. A true automaton—a physical being without any thought processes—could not meaningfully partake in the rights guaranteed by the First Amendment. Thus, some kind of freedom of thought and the right to control one's thought must be a part of the constitutionally protected right of privacy.

This important interest in control of one's mind was recognized by Justice Brandeis in his dissent in an early wiretapping case, *Olmstead v. United States*:

> *The makers of our Constitution undertook to secure conditions favorable to the pursuit of happiness. They recognized the significance of man's spiritual nature, of his feelings and his intellect. They knew that only a part of pain, pleasure, and satisfactions of life are to be found in material things. They sought to protect Americans in their beliefs, their thoughts, their emotions, and their sensations.*[53]

A decade ago in *Stanley v. Georgia* the Supreme Court rejected the suggestion that the state could act to protect the minds of its citizens from the polluting effects of obscenity:

> *Our whole Constitutional heritage rebels at the thought of giving government the power to control men's minds . . . We are not certain that this argument amounts to anything more than the assertion that the State has the right to control the moral contents of a person's thoughts . . . [The State may not] constitutionally premise legislation on the desirability of controlling a person's private thoughts.*[54]

When the Court refused to extend *Stanley* to protect commercial showings of pornography in *Paris Adult Theatre I v. Slaton* it analogized to legal restrictions on the sale of drugs, which it presumed to be valid:

> *[W]e reject the claim that the State of Georgia is here attempting to control the minds or thoughts of those who patronize [these] theatres . . . [T]he fantasies of a drug addict are his own and beyond the reach of government, but government regulation of drug sales is not prohibited by the Constitution.*[55]

The First Amendment derivative theory of privacy has been recognized by two courts. In limiting the authority of a state to impose psychotropic drug therapy upon a patient committed to a state institution, the United States District Court in Massachusetts, in *Rogers v. Okin,* held:

> *The right to produce a thought—or refuse to do so—is as important as the right protected in Roe v. Wade to give birth or abort. Implicit in an individual's right to choose either abortion or birth is an underlying right to think and decide. Without the capacity to think, we merely exist, not function. Realistically, the capacity to think and decide is a fundamental element of freedom.*
> *The First Amendment protects the communication of ideas. That protected right of communication presupposes a capacity to*

> *produce ideas. As a practical matter, therefore, the power to pro-*
> *duce ideas is fundamental to our cherished right to communicate*
> *and is entitled to comparable constitutional protection.*[56]

Similarly, in the celebrated case of *Kaimowitz v. Department of Mental Health* a lower court in Michigan held that "[t]o the extent the First Amendment protects the dissemination of ideas and expressions of thoughts, it equally must protect the individual's right to generate ideas."[57]

The *Rogers* opinion may not go far enough. Not only is the mental process necessary to the communication of ideas protected by the First Amendment, but it is the only reason for the enumerated protections. Thought is not only necessary for meaningful speech, it is the reason for that speech. Thought is not merely a means to an end, it is the end itself.

Even the right to control one's mind is not absolute. Although everything a government does, from designing an agency logo to planning bus routes, has some effect on some people and thus may influence someone's thoughts and decisions, some governmental action affects private individuals' thought in such a *de minimus* way that it could not be held to violate any right to privacy. Other governmental action, like the maintenance of defense secrets, may have a serious effect on individuals' thought, but may still be necessary to serve other Constitutionally mandated interests. Finally, of course, some people may have less of an interest in the maintenance of independent thought, and less of a right to it, than others. For example, those involuntarily committed to a mental institution necessarily give up some of the rights enumerated in the Bill of Rights. As these rights evaporate, so do their penumbras and emanations. As essential as the freedom of thought is to most people, it is less meaningful to one incarcerated and subjected to involuntary care because his inability to think rationally makes him a danger to himself or others.

It is not surprising that "freedom of thought" was not specifically included in the Bill of Rights. Mind altering drugs provided little social problem in the 18th Century, and thought control could be effected only through the control of the input and the ouput of the mind, not the mental process itself. That history should not obscure the logical necessity that this right extend to the mental process.

Control of the Body

Courts have consistently recognized the privacy interest in the control of one's own body. This was, after all, at the center of *Roe v. Wade*, and it has been the subject of the subsequent abortion cases. Other cases that have supported the right of patients to control their medical care have been based on this same kind of fundamental right.[58] One extreme invasion of the right to control one's body is slavery, which is specifically outlawed by the Thirteenth Amendment. The political focus of the social movements that gave rise to the Abolitionist Movement and the Thirteenth Amendment also formed the basis for the Fourteenth Amendment and its due process clause, and it would be strange indeed to read the notions of autonomy, individuality, freedom, and personal control, which were such vital parts of that movement, out of the Fourteenth Amendment.

In any case, the relationship of one's identity to the right to control one's own body is an obvious one. At least some degree of physical autonomy is clearly essential to the enjoyment of other rights guaranteed in the Constitution. As studies of the uses of torture demonstrate, the alienation of one from one's own body is an important step in the breakdown of one's identity as a person.

The Relationship Between Control of Mind and Control of Body

The use of torture provides a good example of the relationship between the mind and the body. Losing complete control of one's body makes it difficult to maintain control of one's mind. Loss of identity invariably involves impingements upon body and thought. This is also demonstrated by the "mortification of self" that Erving Goffman tells us is a consequence of a total institution's ability to cause an individual to lose control of "body, immediate actions, thoughts, and possessions."[59] Thus, one prerequisite to the enjoyment of other constitutionally protected rights is an independent mind in an independent body. Freedom of thought and the freedom to control one's own body are so fundamental to the enjoyment of other constitutional rights that they are necessarily implicit in our concept of ordered liberty.

General Limitations on the Control of Mind and Body

One need not exercise complete and unfettered control of one's body (and, as indicated above, even of one's mind) at every moment to maintain individuality and identity. Although some attributes of these rights are so basic that they must be enjoyed everywhere and at every time—we will not tolerate slavery, even for a moment, anywhere—others are necessary only occasionally, or in particular circumstances. For example, the right to engage in sexual conduct may be essential to our notion of privacy, but it is not so fundamental that one needs access to it in every place and at every time. Thus, the state may limit such conduct to certain circumstances (for example, within a family), and provide it greater protection in certain places (for example, in such private preserves as the home).

Indeed, most of the attributes of privacy can probably be restricted under some circumstances without destroying the individuality associated with them, as long as there is some place where one maintains absolute authority over every attribute of one's mind and body. In *Griswold*, after all, it was the possibility of police invasion of the private precincts of the marital chamber that led to the development of the modern right of privacy.[60] In *Stanley* the United States Supreme Court protected the right of an individual to maintain obscene materials in the privacy of his own home, even though he would not be able to carry them upon the streets.[61] The Court there concluded that it would be wholly improper for the state to limit what a person could read or see (and thus think), at least in the sanctuary of his home.[62] Finally, in *Ravin v. State* the Supreme Court of Alaska found that the state statute prohibiting the possession of marijuana was a violation of the due process clause of the Fourteenth Amendment as well as a separate privacy provision of the Alaska constitution, to the extent that it prohibited the possession in the privacy of an individual's home.[63] The court pointed out:

> *If there is any area of human activity to which a right to privacy pertains more than any others, it is the home. The importance of the home has been amply demonstrated in constitutional law.*[64]

What may be fundamental to one society's concept of ordered liberty may not be fundamental to another's. As the *Ravin* case pointed out, the value of individuality and freedom from government intrusion may be of greater significance in Alaska, where many residents have sought refuge from any authority, than it is elsewhere. Similarly, some groups within society may be able to enjoy the enumerated Constitutional rights with less autonomy and privacy. Children, for example, may not be entitled to the same level of protection of an absolute right to control their minds and bodies as are adults. Thus, in *Parham v. J.R.*, the Court applied a very different standard for the forced detention of children in a mental hospital than had ever been applied in the civil commitment of adults.[65] Although the Supreme Court, in *Planned Parenthood of Missouri v. Danforth* and *Bellotti v. Baird*, concluded that a minor shared in the constitutional right of privacy and thus possessed the right to an abortion, the Court clearly held that the child's right is not identical to an adult's right.[66] In *Bellotti*, Justice Powell explained that

> [t]he status of minors under the law is unique in many respects . . . The unique role in our society of the family [requires] that constitutional principles be applied with sensitivity and flexibility to the special needs of parents and children. We have recognized three reasons justifying the conclusion that the constitutional rights of children cannot be equated with those of adults: the peculiar vulnerability of children; their inability to make critical decisions in an informed, mature manner; and the importance of the parental role in child rearing.[67]

It is apparent that some attributes of control of mind and body are the proper subect of some kinds of government regulation. The question cannot be whether privacy, or personhood, or individuality is "fundamental to our concept of ordered liberty." The answer to that is obviously yes. Similarly, the question ought not be whether the use of amphetamines on Thursday night is "fundamental to our concept of ordered liberty." The answer to that is obviously no. Thus, at least one of the questions that arises out of this evaluation of the right of privacy and the constitutional protection to be afforded it is the nature of the specificity of the questions we must ask.

The Current Constitutional Test for Evaluating Regulation of Fundamental Rights

In evaluating challenges under the due process clause of the Fourteenth Amendment, the Supreme Court has adopted a test that it has employed in resolving equal protection problems. In determining whether a legislative or administrative regulation violates the equal protection clause of the Fourteenth Amendment, the Supreme Court has first asked whether the interest implicated is a fundamental one, or whether the distinction that the government has made constitutes a "suspect classification," that is, one that is likely to impose a burden on a class traditionally subject to discrimination that is not benign.[68] In the absence of a fundamental interest or a suspect classification, the regulation must be merely rationally related to some permissible state end.[69] When this test is applied, regulatory schemes are almost universally found to be constitutionally sound.

Where a fundamental right is at stake, the court will uphold a restriction upon it only upon the showing that there is a "compelling state interest" that cannot be satisfied by any more narrowly drawn, less restrictive alternative.[70] Very few state regulations have survived this "strict scrutiny." Thus, the denomination of the right in question (the "right of privacy" or the "right to use amphetamines on Thursday night") will determine whether a fundamental right is implicated and thus whether the "compelling state interest" analysis is triggered. If the right to use marijuana, amphetamines, or laetrile is a fundamental right, it will be difficult to find any "compelling state interest" to overcome that right; if the use of those drugs is not a "fundamental right" it will not be difficult to discover some rational relationship between the regulation and some permissible state end.

Judicial deference to regulation that does not impinge upon a fundamental interest is broadened by the doctrine developed in *United States v. Carolene Products,* which requires that a court sustain legislation based on unsettled questions of fact, at least until the invalidity of the premises is no longer contested.[71] Thus, any Fourteenth Amendment challenge to a regulation of drugs that does not constitute an impingement on a "fundamental right" will be unsuccessful as long as there is some factual evidence, even if it is very weak evidence, that supports the rationality of the legislature's actions.

Interests Asserted by Those Who Wish Access to Drugs

Although the reason one seeks to exercise a fundamental right is legally irrelevant to the application of the right, it may be helpful to evaluate the interests of those who do seek access to drugs. Those who seek to use pleasure-enhancing drugs do so, by and large, to exercise control over their mind and their thought processes. Just as particular thoughts, particular analyses, and particular mental processes may be the consequences of the ideas digested (through discussion, through reading, or in some other way), mental processes may be induced by the use of hallucinogens or other kinds of drugs. One using such a drug may assert that he or she has as much right to shape his or her own mental processes through the substances one ingests as to shape those processes through one's choice of reading materials.

Those who use performance-enhancing drugs do so most often to control the reaction of their body. Of course, some "performance-enhancing" drugs may have the effect of causing their user to believe that the body is acting more effectively, even when it is not. Such drug use is really indistinguishable from the use of the pleasure-enhancing drug. One who uses a performance-enhancing drug to control the reaction of the body will assert that the use of the drug is no different from the use of exercise, diet, or proper medical care, all of which are surely protected by the right of privacy.

Interests Asserted by the State in Controlling Access to Drugs

A wide range of interests have been proferred to justify the regulation and prohibition of drug use. As the legal analysis indicates, it is appropriate to determine which of these interests are rational ones and, to the extent that the right to use drugs is a fundamental right, which of these reasons may be "compelling." Generally, the interests asserted by the state fall into two categories: nonpaternalistic and paternalisitic.

Nonpaternalistic Interests

Several nonpaternalistic interests have been urged by states in support of their intrusions into grounds that may be protected by the constitutional right of privacy. The most often asserted interest is the maintenance of public health. For example, a state may wish to regulate some drugs because driving under their influence will cause a risk to the community at large, or because those drugs cause their users to become violent or sexually unrestrained, as was alleged years ago in the case of marijuana. Another reason for regulating access to drugs is the need to overcome fraud, which the state believes is inherent in the nature of any commercial drug transaction. This method of overcoming fraud provides the Food and Drug Administration's primary justification for its dogged attempts to wipe out the use of laetrile.[72]

A third nonpaternalistic state interest in controlling drugs is the protection of the public treasury. Drug users may cease to contribute to the economy, become public charges, and thus a financial burden upon the community. This "public fisc" justification has been used to uphold statutes requiring motorcycle helmets (where, several courts have found, those without the helmets could easily become patients at public hospitals)[73] and to overcome individuals' decisions to forego medical treatment (where, for example, the refusal to undergo a blood transfusion could make the patient and her family public charges).[74] Unfortunately, as Professor Tribe points out, every kind of regulation can be justified in this way.[75] As long as the state maintains social services for the indigent, any activity that can ever result in poverty is subject to public prohibition. This same argument could be used to control the kind of occupation one might enter, to prohibit retirement, and to outlaw skiing. If the state's interest in protecting the public treasury is a compelling one, there is nothing left of the constitutional right of privacy.

Other interests that could be asserted by the state to justify the regulation or prohibition of certain kinds of drugs include esthetics, public tranquility, and the social need to "maintain the fabric of society." For example, the esthetic interest in maintaining "natural" athletic competition unaugmented by drug-induced behavior could justify a state's determination to ban certain performance-enhancing drugs, at least under many circumstances.

Finally, statutes limiting access to pornography and outlawing certain kinds of sexual conduct, notably homosexuality and sodomy, have been justified because of their necessity to maintain the fabric of society.[77] A justification for such a state interest is really very much like the justification for the right of privacy generally—there are some kinds of conduct so traditionally abhorred that their prohibition is "fundamental to our concept of ordered liberty." As is the case in justifications based upon the protection of the public treasury, state justifications based upon esthetics, public tranquility, or the maintenance of the fabric of society can be used to justify essentially every limitation on what would otherwise be encompassed by the right of privacy.

Paternalistic Interests

The state may also justify its intervention in a person's determination of whether or not to use drugs on entirely paternalistic grounds. Most commonly, drug laws are based on the preservation of the mental and physical health of those who would otherwise use those drugs.[78] In addition, the state's intervention to prohibit drugs under some circumstances (for example, the prohibition of the use of drugs in athletic enterprises) may relieve a competitor from the burden of choosing whether or not to use those drugs. Neither athletic competition nor the health of the competitor may be advanced by the existence of this choice.

Yet another paternalistic justification for the state intervention to limit these drugs might be found within our description of the basis of the right of privacy itself. If the maintenance of mental activity—of rational thinking—is fundamental to individuality and personhood, and thus necessary for the enjoyment of all other constitutional rights, might not the preservation of that activity itself be a reasonable endeavor on the part of the state? Under such an analysis, the state may have an affirmative obligation to foster rational thought. Although, of course, the state could not intervene to effect the content or the substance of thoughts, perhaps it is obliged to further the activity generally.[79]

Such an analysis is very much like the view of the religion clauses of the First Amendment associated with Roger Williams. Although he believed that it would be improper for the state to take a position as to the content of religion, or as among religions, he believed the state had an obligation to "countenance, encourage and supply" all religions.[80] Indeed, rational thought may serve the same function in the Twentieth Century that religious thought served in the Seventeenth and Eighteenth Centuries. Just as the capacity for religious belief was presumed to be the unique province of human beings, today rational thought is believed to distinguish human beings from other species. It is this capacity that gives value to human personality and autonomy. Although the state interest in maintaining mental activity may be more obviously related to the fundamental right of privacy, a state interest in maintaining the physical individuality and the physical self-control of those within its jurisdiction may also further the state's interest in permitting individuals to seek the benefits more specifically enumerated in the Constitution. Young girls' use of anabolic steroids to improve their athletic prowess, for example, may result in such unusual physical development that the athletes will be unable to participate in adult activities as women.

Whatever the value of the state's nonpaternalistic interests in controlling drugs, any essentially paternalistic interests seem at odds with a notion of privacy based on individual mental and physical autonomy. But although paternalistic state interests may not generally be able to overcome claims based on privacy, the state's paternalistic interests may be sufficient in particular cases where government paternalism has traditionally been recognized. Most obviously, the social interest in protecting children from their own folly has long been recognized. Even where states have abandoned motorcycle helmet requirements, they have maintained them for minors.[81] Similarly, obscene material available to adults may not be made available to children, and alcohol is available only to adults. Though *Planned Parenthood of Missouri v. Danforth, Bellotti v. Baird,* and *H. L. v. Matheson* all recognized that the right of privacy does extend to children, they also recognize that the right must be modified by the special nature of childhood and the traditionally recognized family relationships.[82] The mere fact that children share in the right of privacy to seek an abortion or contraceptives does not mean that they share in every aspect of the right otherwise extended to adults.

Maintaining the mental and physical health of children has always been within the scope of the state authority. In addition, to the extent that the maintenance of mental activity is a legitimate paternalistic state interest, it may be much stronger among children, whose identity may be more easily shattered by the use of drugs. For example, adolescents may be less able to use some kinds of drugs and also maintain control of their own thought processes than are adults, and the maintenance of the physical characteristics of young girls may be far more important to the development of their personality than is the maintenance of those characteristics in women.

The First Amendment Speech and Religion Clauses, the Equal Protection Clause, and Drug Culture

The right of privacy is not the only constitutional basis for attacking state regulation or prohibition of drug use. Although a discussion of alternative constitutional attacks on drug regulation is beyond the scope of this presentation, there are three arguments that deserve mention because of their close relationship to arguments based on the right of privacy. First, state statutes and federal regulations that prohibit the use of drugs central to religious activity have been attacked under the free exercise clause of the First Amendment.[83] Because the Supreme Court has recognized that the free exercise clause protects religious thought and not conduct, the courts could enforce drug statutes consistently with First Amendment requirements.[84] However, the notion that religious conduct can be separated from religious thought is obviously a very weak one. Thus, the courts have upheld the use of peyote by members of the Native American Church, but only because the use of peyote is central to the religious ceremony itself, because the religion and the ceremony were within a well-defined tradition (and thus, apparently, respectable), and because the religion admitted as adherents only native Americans.[85] The fear of abuse of the free exercise clause of the First Amendment—that is, the fear that some might establish a religion that "required" drug use so as to avoid statutory and regulatory restrictions—was simply insignificant in the case of the Native American Church. On the other hand, in *United States v. Kuch* the District of Columbia District Court concluded that all state and federal regulations limiting drug use applied even to the ceremonies of the Neo American Church.[86] That church, symbolized by its three-eyed toad logo and its "victory over horseshit" motto, provided the court with a clear example of an attempt to use the free exercise clause for nonreligious purposes.[87]

There also may be a First Amendment free speech objection to drug regulation, at least in the case of the users of some kinds of drugs. For example, the users of marijuana may claim that the public use of that drug is part of an expression of their life style, and that such use has specific political meaning. The First Amendment protects communicative conduct as well as free speech.[88] In fact, the history of the control of marijuana may make more apparent the political nature of the drug. When the drug was used by a tiny minority of blacks, who themselves were the subject of severe discrimination, the possession and use of the drug were serious felonies. As the use of the drug spread to the children of the elite, penalties for use and possession declined substantially. Now that those children are themselves coming into positions of power, it is hard to find a community that has not effectively decriminalized personal use of the drug, or at least seriously considered doing so. As it becomes apparent that marijuana is no more harmful than the socially acceptable alcohol, and that the two drugs have been used by entirely different social groups, the prohibition appears even more political.

This same kind of argument has been advanced in other areas where claims to the right of privacy have also been at stake. For example, Professor Tribe suggests that state statutes requiring that motorcyclists wear helmets are probably directed primarily at those motorcycle gangs that appear so menacing on the highways.[89] To a member of the Hell's Angels, clothing (and helmet gear) are obviously a social statement. The political nature of motorcycle helmet statutes is supported by the fact that while many states have promulgated such statutes, no states have imposed a requirement that seatbelts be worn at all times on the highways, even though such a requirement would save far more lives than does the law requiring that motorcyclists wear helmets.

Finally, some drug regulations may be attacked on the grounds that they violate the equal protection clause of the Fourteenth Amendment. This clause requires that any regulatory distinction that does *not* implicate a fundamental right must, nonetheless be a rational distinction. Since every state drug regulation or prohibition that leaves some drug unregulated or prohibited makes a distinction, every such regulation or prohibition is subject to evaluation under the equal protection clause. Although permitting access to alcohol but not marijuana, or permitting access to caffeine but not amphetamines, may not be rational, arguments suggesting this have uniformly been unsuccessful. The Supreme Court has consistently held that a state is not required to address every part of a social problem that it attempts to resolve.[90] Since there would be severe social and political problems in any attempt to further limit access to alcohol or caffeine, it seems reasonable for the state to focus its attempts to limit drug abuse on drugs with uses more easily contained.

In any case, it is unlikely that any equal protection claims based on the fact that a drug is regulated in a manner unlike alcohol will ever succeed. The Twenty-first Amendment, promulgated to repeal prohibition, has been interpreted to give plenary authority to the state legislatures to control alcohol.[91] Thus, in pursuit of the control of alcohol, the states may apply a regulatory scheme that would be impermissible in dealing with any other drug.[92]

Conclusion

The development of the Constitutional right of privacy does not dictate a model for federal and state prohibition and regulation of drug abuse. Although the problems inherent in any wholly legal regulation of the use of drugs ought to cause us to look to other kinds of social controls, the Constitutional analysis does instruct us that any regulatory scheme ought to be the consequence of careful evaluations of precisely what kinds of drugs are to be regulated for what purposes, among what populations, and under what circumstances. Our conclusions ought to be informed by the deference we believe the society should pay to individuals' determinations of how they will control their thought processes and physical development. There is no reason to believe that any single regulatory scheme can be applied appropriately to all kinds of drugs and under all circumstances.

Even if drug use were to be construed to be within the Constitutional right of privacy, that would not mean that such use must remain unregulated. Just as commercial speech such as advertising, although protected by the First Amendment guarantee of free speech, can be made the subject of reasonable "time, place and manner" restrictions, the use of drugs surely can be subjected to such reasonable limitations. For example, the right of privacy has greater breadth in the private home than outside of it. Restrictions on conduct otherwise protected by the right of privacy have been upheld in pornography, sexual conduct, and in one state, marijuana cases.

The law clearly provides that the nature of the right of privacy extended to children is different from the nature of that right enjoyed by adults. Thus, to the extent that a child's autonomy and personality are less well developed than an adult's, so the right of privacy enjoyed by that child will be narrower.

The judiciary has not ended the debate on the prohibition and regulation of drug use. Although judicial discussions of the right of privacy may have contributed to furthering that debate, the Supreme Court's denial of certiorari in the *Rutherford* and *Privitera* cases may demonstrate that the Court recognizes that other fora, such as the family, social groups, and the legislature, may be better able to deal with the many social policy considerations that must underlie the regulation of drug use.

Acknowledgments

The author gratefully acknowledges the help provided by Thomas Murray, Robert Kinscherff, Venessa Merton, Corinne Henning-Sachs, and Raul, whose lively discussion of these issues helped illuminate them for the author.

References and Notes

[1] See Corwin, The "Higher Law" Background of American Constitutional Law, 42 Harv. L. Rev. 149 (1928); Gerety, Redefining Privacy, 12 Harv. Civ. Rts.–Civ. Lib. L. Rev. 233 (1977); Reiman, Privacy, Inti-

macy, and Personhood, 6 Phil. & Pub. Aff. 26 (1976); Thomson, The Right to Privacy, 4 Phil. & Pub. Aff. 295 (1975); Warren and Brandeis, The Right to Privacy, 4 Harv. L. Rev. 193 (1890); Comment, A Taxonomy of Privacy: Repose, Sanctuary, and Intimate Decision, 64 Cal. L. Rev. 1447 (1976); Gavison, Privacy and the Limits of Law, 89 Yale L. J. 421 (1980); Westin, *Privacy and Freedom* (1967); Glancy, The Invention of the Right to Privacy, 21 Ariz. L. Rev. 1 (1979).

[2]Warren and Brandeis, The Right to Privacy, 4 Harv. L. Rev. 193 (1890).

[3]*Id.* at 205.

[4]*Union Pacific Railway Co. v. Botsford*, 141 U.S. 250, 251 (1891).

[5]US Constitution, Amendment XIV.

[6]*Meyer v. Nebraska*, 262 US 390, 399 (1923).

[7]*Id.* at 402.

[8]*Pierce v. Society of Sisters*, 268 US 510 (1925).

[9]*Pierce v. Society of Sisters*, 268 US 510, 535 (1925); *Meyer v. Nebraska*, 262 US 390, 400 (1923).

[10]See *Lochner v. New York*, 198 US 45 (1905).

[11]See *Adkins v. Childrens Hospital*, 261 US 525 (1913).

[12]See *Coppage v. Kansas*, 236 US 1 (1915).

[13]See Gunther, *Constitutional Law* 502 (1980).

[14]*Griswold v. Connecticut*, 381 US 479 (1965).

[15]*Id.* at 484.

[16]*Id.* at 484.

[17]*Id.* at 485–486.

[18]*Id.* at 486.

[19]*Id.* at 488. Justice Goldberg wrote the concurring opinion, in which Chief Justice Warren and Justice Brennan joined.

[20]*Id.* at 499–502.

[21]The phrase was first used by Justice Cardozo to define that which was included within the due process clause of the Fourteenth Amendment in *Palko v. Connecticut*, 302 US 319, 325 (1937).

[22]*Poe v. Ullman*, 367 US 497 (1961) (Harlan, J., dissenting).

[23]*Id.* at 542.

[24]*Griswold v. Connecticut*, 381 US 479, 510 (1965) (Black, J., dissenting).

[25]*Eisenstadt v. Baird*, 405 US 438 (1972).

[26]*Id.* at 453.

[27]*Stanley v. Georgia*, 394 US 557 (1969).

[28]*Doe v. Bolton*, 410 US 179 (1973); *Roe v. Wade*, 410 US 113 (1973).

[29]*Roe v. Wade*, 410 US 113, 152 (1973).

[30]*Id.* at 159.

[31]*Doe v. Bolton*, 410 US 179, 211 (1973) (Douglas, J., concurring).

[32]*Id.*

[33]*Id.* at 213.

[34]The Court recently reaffirmed *Roe v. Wade* in *City of Akron v. Akron Center for Reproductive Health, Inc.*, 51 USLW 4767 (US, June 15, 1983). Congress had attempted to exert its power over abortion. One "Right to Life" Amendment, SJR 3, failed passage 49 to 50 in June, 1983. Other Right to Life Amendments have been introduced (SJR 8, 9) and are awaiting action.

[35]*See, e.g., H.L. v. Matheson*, 450 US 393 (1981); *Bellotti v. Baird*, 443 US 622 (1979); *Planned Parenthood v. Danforth*, 428 US 52 (1976).

[36]*Harris v. McRae*, 448 US 297 (1980); *Beal v. Doe*, 432 US 438 (1977); *Maher v. Roe*, 432 US 464 (1977); *Poelker v. Doe*, 432 US 519 (1977).

[37]*Whalen v. Roe*, 429 US 589 (1977).

[38]*Id.* at 594.

[39]*Id.* at 600.

[40]*Id.*

[41]*Id.* at 603.

[42]*United States v. Rutherford*, 422 US 544 (1979). Subsequently, at least one federal court has upheld a challenge to a statute designed to deny unconventional medical treatment to those seeking it. In *Andrews v. Ballard*, 498 F. Supp 1038 (SD, Tex. 1980), the court held that a patient had the right to obtain accupuncture from other than a physician, contrary to the Texas Medical Practice Act. *See also*, Laissez Faire in the Medical Marketplace—Recognition of a Constitutional Right to Unconventional Medical Treatment: Andrews v. Ballard, 18 New Eng. L. Rev. 149 (1982); The Uncertain Application of the Right of Privacy in Personal Medical Decisions: The Laetrile Cases, 42 Ohio St. L. J. 523 (1981); Ainsworth and Hall, Laetrile: May the State Intervene on Behalf of a Minor? 30 U. Kan. L. Rev. 409 (1982); and Volzer, Laetrile and The Privacy Right in Decisional Responsibility, 26 Med. Trial Tech. Q. 395 (1980).

[43]*Rutherford v. United States*, 438 F. Supp. 1287 (WD, Okla. 1977).

[44]*Rutherford v. United States*, 582 F.2d 1234 (10th Cir. 1978).

[45]*United States v. Rutherford*, 442 US 544 (1979).

[46]*Rutherford v. United States*, 616 F.2d 455 (10th Cir. 1980).

[47]*Rutherford v. United States*, 449 US 937 (1980).

[48]*Privitera v. California*, 444 US 949 (1979).

[49]*People v. Privitera*, 74 Cal. App. 3d 936, 141 Cal. Rptr. 764 (1978).

[50]*Id.* at 784.

[51]*People v. Privitera*, 23 Cal. 3d 697, 591 P.2d 919, 153 Cal. Rptr. 431 (1979).

[52]*See, e. g.*, Meulders-Klein, The Right Over One's Own Body: It s Scope and Limits in Comparative Law, 6 B.C. Int'l. & Comp. L. Rev. 29 (1983); Goldberg, "Interpretations" of "Due Process of Law"—A Study in Futility, 13 Pac. L. J. 365 (1982); Dworkin, "Natural" Law Revisited, 34 U. Fla. L. Rev. 165 (1982); Ross, A Natural Rights Basis for Substantive Due Process of Law in US Jurisprudence, 2 Univ. Human Rights 61 (1980); and Gavison, Positivism and the Limits of Jurisprudence: A Modern Round, 91 Yale L. J. 1250 (1982).

[53]*Olmstead v. United States*, 277 US 438, 478 (1928).

[54]*Stanley v. Georgia*, 394 US 557, 565–566 (1969).

[55]*Paris Adult Theatre I v. Slaton*, 413 US 49, 67–68 (1973).

[56]*Rogers v. Okin*, 478 F. Supp. 1342, 1367 (D. Mass. 1979). For subsequent history of this case, see fn. 58.

[57]*Kaimowitz v. Department of Mental Health*, Civ. No. 73–19434-AW (Mich. Cir. Ct. Wayne Cty., July 10, 1973).

[58]*Rogers v. Okin*, 478 F. Supp. 1342, 1367 (D. Mass. 1979). The Court of Appeals criticized the district court for its simplistic approach, but generally agreed that drugs should not be forceably administered except under certain conditions. The Court of Appeals found that the right to refuse antipsychotic drugs had its origins in the due process clause of the Fourteenth Amendment. 634 F.2d 650 (1st Cir. 1980). The Supreme Court vacated the Court of Appeals decision for essentially procedural reasons in *Mills v. Rogers*, 102 S. Ct. 2442 (1982). The Supreme Court questioned whether the state's protection of the individual's interest was broader than the protection afforded by the United States Constitution. *See also Rennie v. Klein*, 653 F.2d 836 (3rd Cir. 1981), A Mental Patient's Right to Refuse Antipsychotic Drugs: A Constitutional Right Needing Protection, 57 Notre Dame L. Rev. 406 (1981); and Symonds, Mental Patients' Right to Refuse Drugs: Involuntary Medication as Cruel and Unusual Punishment, 7 Hastings Const. L.Q. 701 (1980).

[59]E. Goffman, *Asylums* (1961).

[60]*Griswold v. Connecticut*, 381 US 479 (1965).

[61]*Stanley v. Georgia*, 394 US 557 (1969).

[62]*Id.* at 565.

[63]*Ravin v. State*, 537 P.2d 494 (Alaska 1975).

[64]*Id.* at 503.

[65]*Parham v. J. R.*, 442 US 584 (1979).

[66]*H. L. v. Matheson*, 450 US 393 (1981); *Bellotti v. Baird*, 443 US 622 (1979); *Planned Parenthood v. Danforth*, 428 US 52 (1976).

[67]*Bellotti v. Baird*, 443 US 622, 633–34 (1979).

[68]Gunther, *Constitutional Law* 671 (1980).

[69]*Id.* at 670–671.

[70]*Id.* at 671.

[71]*United States v. Carolene Products Co.*, 304 US 144 (1938).

[72]*See* Laetrile: The Making of a Myth, HEW Pub. No. (FDA) 77-3031; Toxicity of Laetrile, FDA Drug Bulletin 7 (5) (Nov.–Dec. 1977).

[73]*See* e.g., *State v. Craig*, 19 Ohio App. 2d 29, 249 N.E. 2d 75 (1969); *State v. Odegaard*, 165 N.W. 2d 677 (N.D. 1969); *State v. Albertson*, 93 Idaho 64, 470 P.2d 300 (1970); *State v. Lombard*, 241 A.2d 625 (R.I. 1968); and *Simon v. Sargent*, 346 F. Supp. 277 (D. Mass. 1972), *aff'd mem.*, 409 US 1020 (1972). The Ohio helmet law upheld in *Craig* was repealed in part in 1978 to eliminate the helmet requirement except for those under 18. For an examination of helmet laws, *see* Helmetless Motorcyclists—Easy Riders Facing Hard Facts: The Rise of the "Motorcycle Helmet Defense," 41 Ohio St. L. J. 233 (1980).

[74]*See*, e.g., *Ex Parte Hilley*, 405 So. 2d 708 (Ala. 1981); *In Re Lucille Boyd*, 403 A.2d 744 (DC Ct. App. 1979); and *Hamilton v. McAuliffe*, 277 Md. 336, 353 A.2d 634 (Ct. App. 1976).

[75]Tribe, *American Constitutional Law* (1978), 938–941.

[76]*See* note 73, *supra*.

[77]*See*, e.g., *Manual Enterprises v. Day*, 370 US 478 (1962); *Hearn v. Short*, 327 F. Supp. 33 (S.D. Texas 1971); and *Auginblick v. US*, 206 Ct. Cl. 74, 509 F. 2d 1157 (1975). *See also, Commonwealth v. Bonadio*, 490 Pa. 91, 415 A.2d 47 (1980); *People v. Onofre*, 51 NY 2d 476, 415 N.E. 2d 936 (1980), *cert. denied*, 101 S. Ct. 2323 (1981); Commonwealth v. Bonadio: Voluntary Deviate Sexual Intercourse—A Comparative Analysis, 43 U. Pitt. L. Rev. 253 (1981); and Sexual Morality and the Constitution: People v. Onofre, 46 Albany L. Rev. (1981). In *Onofre* and *Bonadio*, the courts struck down sodomy statutes as unconstitutional.

[78]*See*, e.g., H. R. Rep. No. 2464, 87th Cong., 2d Sess., enumerating the purposes of the Drug Amendments of 1962.

[79]This argument has also been advanced to support deprogramming of those in religious cults. *See*, Delgado, Religious Totalism as Slavery, 9 NYU Rev. Law and Soc. Change 51 (1979).

[80]Quoted in Tribe at 817.

[81]*See* note 73, *supra*.

[82]*Bellotti v. Baird*, 443 US 622 (1979); *Planned Parenthood v. Danforth*, 428 US 52 (1976); *H. L. v. Matheson*, 450 US 393 (1981).

[83]*See, People v. Woody*, 61 Cal. 2d 716, 394 P.2d 813, 40 Cal. Rptr. 69 (1964).

[84]Gunther, *Constitutional Law* 1585 (1980). *See Cantwell v. Connecticut*, 310 US 296 (1940); and *Reynolds v. United States*, 98 US 145 (1878).

[85]*People v. Woody*, 61 Cal. 2d 716, __, 394 P.2d 813, 816–818, 40 Cal. Rptr. 69, 72–74 (1964).

[86]*United States v. Kuch*, 288 F. Supp. 439 (D.D.C. 1968).

[87]*Id.*

[88]In *Thornhill v. Alabama*, 310 US 88 (1940), the Court invalidated on free speech grounds an Alabama law prohibiting picketing. However, the Court has distinguished speech from conduct, and its decisions on "symbolic speech" are inconsistent. Compare *Cox v. Louisiana*, 379 US 559 (1965) with *Edwards v. South Carolina*, 372 US 299 (1963). *See generally*, Tribe, *supra*, at 598–601.

[89]Tribe, *supra*, 940–941.

[90]*See Railway Express Agency v. New York*, 336 US 106 (1949).

[91]*See*, e.g., *California v. La Rue*, 409 US 109 (1972); *Lar kin v. Grendels' Den, Inc.*, 51 USLW 4025 (1982); *Women's Liberation Union of Rhode Island v. Israel*, 379 F. Supp. 44 (D.R.I. 1974).

[92]*California v. La Rue*, 409 US 109 (1972). *But see Craig v. Boren*, 429 US 190 (1976).

Using and Refusing Psychotropic Drugs

Nancy K. Rhoden

Introduction

Proponents of the decriminalization of marijuana have long argued that various constitutional guarantees, including the right to privacy, should protect the personal use of this relatively harmless drug. With a few notable exceptions, our courts have not agreed with this claim.[1] But recently there has been a series of constitutional cases involving individual autonomy and drug use that might possibly expand the right of the individual to make such decisions. Specifically, several federal courts have recently held that patients in state mental hospitals have a constitutional right to refuse psychotropic (mind-altering) drugs.[2]

Although the actual holdings apply only to the facts of the cases, the constitutional theories invoked in each of the cases may provide new support for the argument that persons have a right to use mind-altering substances for their own purposes. Whether the "right to refuse treatment" cases will significantly bolster arguments for decriminalization of marijuana and other "recreational" drugs will depend upon: (1) whether the right to refuse is upheld in the Supreme Court, and/or widely accepted in state or lower federal courts; (2) whether the constitutional theories invoked are applicable to situations other than that of mental patients being forcibly medicated; and (3) whether there are sufficient parallels between a mental patient claiming the right to refuse mind-altering medication and a member of the general populace claiming the right to use a mind-altering substance for nonmedicinal purposes. After describing the refusal

157

of treatment cases and their constitutional underpinnings, this article will explore the similarities and differences in these situations and will argue that the right to refuse treatment cases do tend to bolster arguments for a right to use mind-altering drugs.

The Right to Refuse Psychotropic Drugs

In a series of decisions beginning with *Rennie v. Klein*, federal courts have held, in similar but unrelated cases, that mental patients in state hospitals have a constitutional right to refuse the most widely used therapy today—treatment with psychotropic drugs. Although recognition of this right had been urged by legal commentators and foreshadowed by a number of decisions allowing mental patients to refuse unusually intrusive or hazardous treatments, these decisions nonetheless constitute a far-reaching and controversial expansion of the rights of mental patients. The cases concerning the right to refuse psychotropic medication have all involved large state hospitals where many patients were routinely given medication without their consent. The medications they were given were primarily antipsychotic drugs such as thorazine, which are used to treat schizophrenia and which, though they do not cure the disease, do in many cases reduce the incidence and severity of hallucinations, delusions, and other schizophrenic symptoms. These medications, although more effective than other treatments for schizophrenia, unfortunately all have serious, and sometimes permanent, side effects. Although patients who refused these medications often did so because of their side effects, the facts of the cases indicate that hospital personnel neither adequately monitored the patients' responses to the drugs nor responded to their legitimate complaints. Consequently, some patients were forcibly medicated who exhibited clear contraindications to such treatment or who had already suffered permanent neurological damage from the drugs. In holding that patients have a right to refuse treatment with psychotropic drugs, the courts were understandably influenced by the risks associated with the medication and the evidence of its frequent misuse.

The courts have derived the constitutional right to refuse treatment from several different legal theories that, though closely related, can be divided into three basic categories. The first makes reference to a person's interest in bodily integrity and to the doctrine of informed consent, which requires doctors to secure a patient's knowledgeable consent before a medical procedure can be performed upon him or her. The second relies on the constitutional right to privacy, or autonomy, in making fundamental personal decisions. The third basis is the notion that the First Amendment protection of freedom of expression also encompasses freedom of thought, such that the state must refrain from forcibly influencing a person's mental processes. These theories are clearly interrelated, and in fact one court dealing with the matter has simply held that persons have a Fourteenth Amendment liberty interest in not being involuntarily medicated, which encompasses various concerns, and which is violated if they are forcibly given medication without a prior due process hearing.[3] In order to determine which, if any, of these legal theories could apply to the right to use mind-altering drugs, we must first look closely at each one in the refusal-of-treatment context.

A person's interest in protecting his or her bodily integrity has long been recognized in our legal system. Under common law, doctors or other medical personnel who perform medical procedures upon persons without their consent commit a battery, barring special circumstances (such as the unconscious patient) where it is not possible to obtain consent. More recently, the doctrine of informed consent has been developed; this requires that for a person's consent to a medical procedure to be valid, that person must not have merely agreed to performance of the procedure, but must also have been advised of the potential risks and benefits involved, and of any alternatives to the procedure. Because the involuntary use of psychotropic medication typically involves restraining the patient and forcibly injecting the drug, or else inducing the patient to take the medication by threats of an injection, this is a rather clearcut invasion of the patient's bodily integrity and the right to give consent, knowledgeable or otherwise, to the treatment. (This discussion is not meant to suggest that involuntary treatment is never justified, but merely to elucidate the individual interests that are at stake—there may, of course, be state interests that override even important individual rights.)

It is not only the integrity of a person's body that is paramount in the informed consent doctrine—the doctrine also serves to protect an individual's fundamental right to make his or her own decisions about personal matters such as health care. Theorists have persuasively argued that the ethical foundation of the informed consent doctrine is not some abstract notion of the general good, or even the desire to protect helpless individuals from harm, but is rather respect for individual autonomy and self-determination.[4] Decisions about fundamental personal matters affecting one's life are most properly the province of the individual, even though someone else, such as a doctor, may possess far more information and training. Recognition that respect for individual autonomy is at the heart of the informed consent doctrine illustrates the close relationship between it and the second legal basis for the right to refuse treatment—the constitutional right to privacy.

Privacy as a constitutional right was first recognized in *Griswold v. Connecticut*,[5] a case that invalidated a state statute prohibiting the use of contraceptives. It was subsequently developed in a series of cases involving personal decisions about marriage, the family, procreation, and other fundamental matters.[6] It has often been noted that this constitutional right to privacy invoked in these cases is not what is ordinarily referred to as "privacy." It has little to do with invasions of private homes, papers, or possessions, or the obtaining of personal information. Rather, it is an interest in autonomy, or liberty, or self-determination in certain peculiarly private and fundamental areas, such that we are as free as possible from governmental intervention in these areas. Once an area is identified as being within the scope of the right to privacy, then governmental regulation in the area is justified only if is serves to advance a compelling state interest.

At present, there are no precise guidelines for determining which personal decisions are within the protected zone of the right to privacy. Obviously, not all personal decisions of individuals warrant constitutional protection. In *Roe v. Wade* (the abortion decision), Justice Blackmun stated that the right to privacy applies only to those areas that are "fundamental" or "implicit in the concept of ordered liberty."[7] Various Supreme Court decisions make clear that decisions involving marriage and contraception are sufficiently fundamental to be protected. Justice Douglas, in his cogent concurring opinion in *Roe v. Wade*, suggested two other general areas that should be similarly protected. These are "the control over the development of one's intellect, interests, tastes, and personality," and "the freedom to care for one's health and person, freedom from bodily restraint or compulsion, freedom to walk, stroll, or loaf."[8] Under this analysis, the right to privacy—i.e., the right to autonomy in certain fundamental areas—would protect a mental patient's right to refuse psychotropic drugs both because this is a personal medical decision and because the decision whether to use a mind-altering drug comes within the ambit of decisions involving the expression of one's intellect and personality. Of course, as noted previously, state interests could at times outweigh this right, such as in cases where forcible medication is necessary to prevent harm to the patient or to other persons.

The third legal theory used in the right-to-refuse-treatment cases, which has been relied upon in only one decision thus far,[9] appears to hold great promise for supporting a right to use mind-altering substances. This is the notion that since the First Amendment protects the free communication of ideas, and since the capacity to freely produce ideas is necessary for communication to take place, the First Amendment must likewise protect the freedom to produce ideas. Because psychotropic drugs may "affect and change a patient's mood, attitude and capacity to think," Judge Tauro, in the district court decision in *Rogers v. Okin*, held that their forcible administration violates an individual's First Amendment right to freedom of thought.[10]

The theory that the First Amendment is designed to permit individuals to think freely and to develop their unique personalities is certainly not a new or novel one. However, the notion that courts can rely directly upon a First Amendment right to freedom of thought in cases where restraints of expression are not at issue is a recent innovation.[11] I have argued elsewhere that relying upon the First Amendment in this context is fraught with many difficulties and is not a viable legal theory in the form set forth in Judge Tauro's decision.[12] However, the basic idea behind the theory is a valid, and very important, one. It is simply that a person's mental processes should be inviolable, and should not be subject to involuntary control or influence by the state. In other words, individuals are entitled to autonomy, to as great an extent as possible, over their mental processes and those actions and substances that affect them. Considered in this light, the notion of a First Amendment right to freedom of thought can essentially be thought of as another form of privacy, such that an individual should have the autonomy to, as Douglas put it, control the development of his intel-

lect and personality. Thus, just as the informed consent doctrine and the right of privacy overlap to indicate that individuals should have the right to make fundamental health care decisions for themselves, the right of privacy and the right to
freedom of thought overlap to demonstrate that an individual's
mind, just as his or her body, is sacrosanct, and should be as
free as possible from influence and coercion by the state.

The Right to Use Psychotropic Drugs

To what extent do the legal theories supporting a right to
refuse psychotropic drugs suggest that persons should also be
free to decide to use a mind-altering substance such as marijuana? There are, of course, some important differences between situations involving the right to refuse drugs and those
involving the right to use them. The violation of individual
rights that takes place when the state forcibly injects a mental
patient with mind-altering substances is a singularly striking
sort of violation, because it involves unauthorized invasion of
the body. State prohibition of the use of a substance, on the
other hand, involves no violation of bodily integrity. Because of
this immediately apparent difference between the two situations, it may seem at first glance that the mere prohibition of a
drug simply has nothing relevant to or in common with the situation in which the state ignores the requirement of informed
consent and forces unwanted medication on a patient.

However, although invasion of the body is clearly important in the right-to-refuse cases, its absence in the right-to-use
analysis should not be considered determinative. Suppose the
state developed a technological device that allowed it to influence a mental patient's thoughts in exactly the same way drugs
such as thorazine do, but without any invasion of the body at
all: patients would simply walk past a sign saying "Admitting
Office" and its radiations would render them remarkably doc-
ile. It is doubtful that even this highly sophisticated form of
thought control would pass constitutional muster, because, as
noted previously, the requirement of informed consent is
based not only on notions of bodily integrity, but also on the
right of the individual to make his or her own health care decisions. Thus the right-to-refuse cases do not distinguish be-

tween forcible injection of medication and coercing a patient to take medication orally via threatening an injection. Thus we must consider whether, despite the lack of bodily invasion in the prohibition of recreational drug use, the right to autonomy or independence in making personal health care decisions (as it is found in both the informed consent doctrine and the constitutional right to privacy), and the right to control over one's personality, thoughts and mental processes (whether derived from the right to privacy or the First Amendment), are applicable when the state seeks to prohibit the use of certain drugs.

When the state forces an unwilling mental patient to take mind-altering medications, it is violating the individual's freedom to make personal decisions. The state's action in such cases touches upon two of the areas that Justice Douglas' privacy analysis suggested should be free from state interference—the freedom to make medical decisions, and the freedom to control one's mental processes. In the health care arena, it has been suggested that the right to refuse treatment finds its corollary in the right to use the treatment of one's choice. The few courts that have upheld a cancer patient's right to obtain and use laetrile have referred to cases that uphold the right of a patient to refuse therapy altogether, reasoning that since the right of privacy permits a patient to forego treatment, he should likewise have a right to choose an unconventional course of treatment, so long as it is not actively harmful.[13] If laetrile is totally ineffective, taking it is equivalent to refusing treatment (providing it is not actively harmful), and commentators have strongly argued that: "There is no basis in logic for refusing to recognize the right to choose an unproven remedy while at the same time protecting a patient's right to refuse proven remedies."[14] An increasing number of states have legalized laetrile within their borders.[15] But whether or not substances such as laetrile become widely used for medicinal purposes, the basic argument that since the right of privacy protects patients in refusing therapy, it should likewise give them a right to choose ineffective or unconventional therapies, has much persuasive force. Of course, this *prima facie* right must be weighed against state interests in regulation—for example, in the laetrile situation the state has a strong interest in protecting persons from fraud and in keeping worthless medicines off the market. But however the competing rights and interests are ultimately weighed, in the medical treatment area the right to refuse a proferred therapy and the right to choose an alternative one do appear closely related.

It has recently been argued that under the privacy doctrine, chemotherapy patients should have a right to use marijuana to combat nausea, and glaucoma patients should be allowed to use it to prevent blindness.[16] Some states are beginning to allow persons with these conditions to participate in experimental programs in which they can use marijuana for its medicinal value.[17] The notion that the right to refuse medical treatment and the right to use unconventional treatments are correlative would, of course, be extremely useful to the issue at hand if large numbers of marijuana users were cancer or glaucoma patients. But since most users of marijuana and other recreational drugs are not seeking therapeutic benefits (at least not of a kind currently recognized by medical science), we cannot rely heavily on the constitutional right to make personal health care decisions. (This is not to imply that this right is entirely irrelevant—some prohibited drugs may be used by certain individuals for the sort of performance enhancement or spiritual insight that at least arguably can be considered a health-related purpose.) Instead we must concentrate on the third prong of the refusal of treatment cases—the right to control one's mental processes, as a part of the right to privacy, the First Amendment, or both. It seems clear that the right to choose to take a substance that affects the mental processes is the corollary of the right to choose to forego a psychotropic drug despite its purported benefits. Thus we can conclude that since deciding whether to take a mind-altering substance is an aspect of controlling one's own mental processes, to the extent that the constitution does or should protect a person's right to control his own thought processes, then there should be at least a *prima facie* right to take such substances.

Despite the persuasive force of the notion that the right to privacy and/or freedom of speech include the right to control one's mental processes, and that such control should extend to both refusing and using mind-altering drugs, the constitutional basis for this right is undoubtedly weaker than the basis for the right to make personal health care decisions. In *State v. Renfro*, in upholding criminal sanctions for the use of marijuana, the court noted that: "The Supreme Court has never intimated that freedom of speech attaches to chemical substances which physically affect the workings of the brain, or that the ingestion of such substances involves the reception of 'information or ideas'." [18] And even in *Ravin v. State,* where the Alaska Supreme Court struck down the criminalization of marijuana use in one's home as a violation of the Alaska constitutional protection of privacy, the court relied on locational privacy rather than holding that the right to use marijuana was of a sufficiently important or fundamental nature to invoke the privacy right to make important personal decisions. [19] However, since deciding what substances to ingest is clearly a personal decision that affects one's mental processes, it would seem to be within the ambit of the privacy right, though probably imbued with less importance and urgency than the right to make personal health care decisions.

Of course, even assuming that there is a constitutional right involved, proper state policy does not follow inexorably. The state can infringe on even constitutionally protected rights so long as it has a compelling interest that justifies the infringement. We noted earlier that because there is a right to make personal health care decisions, including the decision to forego treatment, there should be at least a *prima facie* right to choose unconventional treatments such as laetrile. But determining a policy on this issue is much harder than merely noting that an individual right is at stake, because the state has a legitimate interest in protecting persons from fraud and from the harm caused by the promotion and distribution of worthless remedies. These rights and interests must be weighed, and various courts and legislatures have weighed them differently. Similarly, although there should be a *prima facie* constitutional right to take mind-altering substances, the state may have a compelling interest in prohibiting the use of certain substances. When the use of a particular substance by individuals poses a substantial risk to the individual or to other members of society, as is probably the case with the use of drugs such as heroin, then the state's interest in prohibiting it may legitimately override the individual's right to autonomy in such matters. Thus even though the constitutional reasoning in the right-to-refuse cases suggests that there should be a correlative right to take mind-altering substances, the amount of support this gives to proponents of the decriminalization of marijuana or other drugs will vary depending upon one's assessment of the state's interest in prohibiting such substances. It is beyond the scope of this article to assess the various arguments concerning marijuana's effects and risks for the individual user and for the society at large, much less the effects, risks, and social costs of the myriad other psychotropic drugs used for recreational purposes. However, it may be helpful to briefly examine some of the types of paternalistic state interventions and their justifications, and to take note of a few relevant comparisons between the state's interest in prohibiting drugs such as marijuana and its interest in (1) regulating the medicinal use of substances such as laetrile, and (2) forcibly medicating mental patients.

One type of paternalistic intervention by the state in the lives of its citizens is outright prohibition of certain types of conduct. The state has the power to prohibit activities of individuals that cause a substantial risk of harm to others. Thus it is, for example, unlawful to drive above a certain speed or while intoxicated. It is largely on the basis of such a risk of harm to others that prohibition of narcotics such as heroin is typically justified—supporters of prohibition point to the fact that the heroin addict may steal from others to support his habit, or may disrupt society in other ways. Also, since "hard" drugs pose a substantial risk of harm to the user himself, their prohibition is supported by the theory that the state has an interest in protecting its citizens from even self-inflicted harm. Many courts have accepted this sort of state paternalism, while stressing that drug related damage to individuals is likewise harmful to society as a whole.[20] However, a few courts have found, in other contexts, that the state has no legitimate interest in protecting individuals from causing themselves harm, at least when they have knowledgeably assumed the risk.[21] At any rate, even if the state does have a legitimate interest in prohibiting some dangerous conduct solely on the basis of its potential harm to the actor, this is undertaken in a highly inconsistent fashion in our society, since there is no prohibition of automobile racing, sky diving, smoking tobacco, or drinking alcohol. And there is no prohibition of such activities even though it is easy to point to resulting societal harms, such as injuries caused by drunk drivers, and the enormous cost of treating lung cancer victims. Finally, the state cannot prohibit conduct such as loitering that has no adverse effects on the actor or on others,[22] although even this maxim is not always honored, as witnessed by the continuing criminalization, in some states, of homosexual acts between consenting adults.

Another type of governmental intervention is the regulation of activities that are useful and necessary, but which could be harmful in the absence of some regulation. For example, numerous regulations are designed to ensure that food is unadulterated and drugs are relatively safe and effective. The intent behind such regulation is not the outrightly paternalistic (or coercive, if you will) one of preventing persons from taking substances they wish to take. Rather, the state assumes that no one wants unsafe or ineffective medications, but that absent some regulation, producers would advertise and market such substances and consumers would be taken unawares. It is important to note that this consumer protection rationale permits the state to keep even harmless treatments off the market if they are ineffective, since consumers could be harmed by using worthless remedies rather than more effective ones. The effect of this can at times be highly paternalistic, as in the laetrile situation where consumers may, quite knowledgeably, wish to forego the pain of conventional (and only somewhat effective) cancer therapies in favor of the probably ineffective, but less debilitating, use of laetrile. As noted previously, in this case the argument that if persons have a right to refuse treatment, they should have a right to choose ineffective treatment, is a strong one. The rejoinder, however, is also strong, since promotion of worthless remedies is clearly harmful to persons who are influenced to take them, and since relaxation of laws in this area will unavoidably lead to the promotion of worthless remedies.[23]

How does the state's interest in prohibiting marijuana compare with its interest in regulating the medicinal use of substances such as laetrile? One comparison is that while cancer victims may be badly harmed by taking laetrile rather than pursuing a more fruitful course of treatment, there is no similar danger with the recreational use of drugs such as marijuana, since persons do not ordinarily use recreational drugs because of false hopes of a cure for a serious illness. True, persons could use marijuana without knowledge of its potential hazards, but this could be remedied by warning labels, as with tobacco, or regulations on sales, as with alcohol. Since the consumer protection rationale is missing with recreational drugs, it seems that they can only be prohibited if their intentional use causes harm to the user or to others. In the case of drugs such as heroin, the harm caused to the individual users of the drugs and to society may be sufficient to justify its prohibition. However, with less dangerous drugs such as marijuana, prohibition becomes harder to justify. In other words, because of the dangers of promotion of worthless remedies, the state's interest in regulation to protect consumers is at least arguably more compelling, and less coercive, than its interest in prohibiting the use of relatively harmless recreational drugs. If the state's interest in prohibiting the use of marijuana is in fact less compelling than its interest in medicinal drug regulation, then this may well balance out the fact that an individual's interest in using drugs for pleasure seems less compelling than his interest in using them to preserve his health.

The other important comparison is, of course, between the state's interest in forcibly medicating mental patients and its interest in denying recreational drugs to members of the general populace. Whether we view the paternalistic policy of drug prohibition as mere bullying by the state or consider that the state is benevolently protecting persons from their own folly, the state is clearly making a personal decision for the individual, and justifying it at least partially on the ground that it is for the individual's own good. The state's interest in forcibly medicating mental patients is of course likewise paternalistic. But mental patients are presumably more appropriate candidates for paternalistic policies, since their ability to make reasoned judgments is at least somewhat impaired. When we keep the comparison between the two classes of persons— mental patients and presumably rational adults who wish to use marijuana—firmly in mind, the right-to-refuse cases seem to lend a great deal of support to proponents of decriminalization of marijuana and other relatively harmless recreational drugs. In the right-to-refuse cases, courts are upholding the rights of persons seriously ill with mental and emotional disorders to refuse the very medication that could alleviate their emotional distress. Needless to say, at least some mental patients will be harmed by their refusal. Moreover, in at least some cases, the patient's decision will be upheld even though it may be made for completely irrational reasons that the patient would repudiate were he "in his right mind." It is, of course, possible that harm, either physical or psychological, will inure to some users of marijuana. However, such users will have chosen to take this risk, and there is little reason to believe their judgment is less rational than that of mental patients who refuse medication. Since, for that matter, there is little evidence that casual users of the milder recreational drugs are less "sane" or more in need of paternalistic protection than the rest of the population, it seem that if persons in general can take risks such as sky diving or using alcohol, and if the constitution protects the right of mental patients to refuse treatment with psychotropic drugs, it should likewise provide ordinary citi-

citizens with at least a *prima facie* right to take psychotropic substances for recreational purposes.

Conclusion

A number of arguments have been advanced in support of a mental patient's emerging right to refuse psychotropic medication, and these arguments center on the importance of individual autonomy in certain fundamental areas. Forcible medication infringes upon individual decision-making authority in several protected areas—bodily integrity, decisions about one's personal medical care, and decisions about how one's thought processes will be influenced or altered. Although prohibition of recreational drug use involves only one of these protected areas—the privacy right or first amendment right to control one's mental processes—nonetheless the principle that individual autonomy should extend to the decision whether to ingest substances that affect the mind seems applicable to the issue of drug prohibition. The effect of recognizing the extension of this principle is that individuals should have a *prima facie* constitutional right to take mind-altering drugs. The analysis, however, does not end here. Whether the state can legitimately override this right depends upon whether it has a compelling interest in prohibiting the use of a particular drug. This will depend upon a complex analysis of the effects of the drug on the individual users and on other members of society. Thus the constitutional analysis being developed in the right-to-refuse cases, though it does not dictate a policy regarding drug prohibition, does indicate that there should be a *prima facie* right to take mind-altering drugs. The burden is therefore shifted to the state to justify its policy of prohibition and, in the case of relatively harmless drugs such as marijuana, this burden is a rather hard one to meet.

References and Notes

[1]See Ravin v. State, 537 P.2d 494 (Alas. 1975), holding that the Alaska constitutional right to privacy in the home is violated by criminalization of private use of marijuana.

[2]See, e.g., Davis v. Hubbard, 506 F. Supp. 915 (D. Ohio 1980); Rogers v. Okin, 478 F. Supp. 1342 (D. Mass. 1979), aff'd 634 F.2d 650 (1st Cir. 1980), remanded for consideration of state law, 457 U.S. 291 (1982); Rennie v. Klein, 653 F. 2d 836 (3d Cir. 1981), remanded for further consideration, 102 S. Ct. 3506 (1982).

[3]Davis v. Hubbard, 506 F. Supp. 915.

[4]See Beauchamp & Childress, *Principles of Biomedical Ethics* 63 (1979); Veatch, "Three Theories of Informed Consent," in *The Belmont Report,* Appendix Vol. II, at 26-19 (National Commission for the Protection of Human Subjects of Biomedical and Behavioral Research, 1978).

[5]381 US 479 (1965)

[6]See, e.g., Loving v. Virginia, 388 US 1 (1967); Roe v. Wade, 410 US 113 (1973); Carey v. Population Services International, 431 US 678 (1977).

[7]410 US 113, 152–53.

[8]410 US 113, 213.

[9]Rogers v. Okin, 478 F. Supp. 1342. The circuit court based its affirmance entirely on the right to privacy, and declined to consider whether protected First Amendment interests were involved. The first amendment rationale had previously been enunciated in Kaimowitz v. Dept. of Mental Health, 2 Prison Law Rptr. 433 (1973), a case involving psychosurgery.

[10]478 F. Supp. at 1366–67.

[11]This theory was originally promulgated by M. Shapiro in "The Use of Behavior Control Technologies: A Response," 7 *Issues in Criminology* 55 (1972) and later elaborated upon in "Legislating the Control of Behavior Control: Autonomy and the Coercive Use of Organic Therapies, 47 *S. Cal. L. Rev.* 237 (1974).

[12]See Rhoden, "The Right to Refuse Psychotropic Drugs," 15 *Harv. Civ. Rts–Civ. Libs. L. Rev.* 363 (1980).

[13]See Rutherford v. United States, 438 F. Supp. 1287, 1298–1300 (W.D. Okla. 1977), aff'd 582 F.2d 1234 (10th Cir. 1978). After a complicated procedural history, this decision was ultimately reversed. 442 US 544 (1979). See also Carnohan v. United States, 616 F.2d 1120 (9th Cir. 1980); Rizzo v. United States, 432 F. Supp. 356 (EDNY 1977).

[14]Note, "The Right to Choose an Unproven Method of Treatment," 13 *Loyola L.A. L. Rev.* 227, 234 (1979).

[15]See Comment, "The Uncertain Application of the Right of Privacy in Personal Medical Decisions: The Laetrile Cases," 42 *Ohio St. L. J.* 523, 530 (1981).

[16]See, e.g., United States v. Randall, 104 Daily Wash. L. Rptr. at 2249. See generally "State Interference with Personhood: The Privacy Right, Necessity Defense and Proscribed Medical Therapies," 10 *Pacific L.J.* 773 (1979).

[17]See "State Interference," *supra* note 16, at 795.

[18]542 P.2d 366, 369 (1975).

[19]537 P.2d 494, 502.

[20]See Borras v. State, 229 So.2d 244 (Fla. 1969), upholding marijuana prohibition on the grounds that the state has a valid interest in having healthy citizens, and that marijuana use is a threat to society in general as well as being harmful to individuals.

[21]See, e.g., American Motorcycle Ass'n v. Davids, 11 Mich. App. 351, 158 N.W.2d 72 (1968), cert. denied 393 U.S. 1037 (1969) (overturning motorcycle helmet requirement).

[22]Papachristou v. City of Jacksonville, 405 US 156 (1972).

[23]See Comment, "Picking Your Poison: The Drug Efficacy Requirement and the Right of Privacy," 25 *UCLA L. Rev.* 577, 613–15 (1978).

Part IV

Drugs, Models, and Moral Principles

Doctors, Drugs Used for Pleasure and Performance, and the Medical Model

Robert Michels

Medical Control of Drugs

Doctors prescribe drugs to make their sick patients well. Everybody knows that this is so, and most people approve of it, at least most of the time. However, there are some exceptions, some problems, and even some contradictions to this general formula. The exploration of these may influence our views on the use of drugs for pleasure and for the enhancement of performance.

At times drugs are administered or prescribed by health personnel who are not physicians. Anesthetists and optometrists administer drugs that are not available to the general public, and pharmacists are frequently consulted regarding the many drugs that are available, but not generally known. People who are neither sick nor patients use drugs. Pain, grief, insomnia, and anxiety all occur in the absence of disease or pathology, and yet are often "treated" by drugs.

Drugs are used to prevent gastroenteritis, motion sickness, or conception, thereby reducing the risk of undesirable complications to various popular activities. Drugs are used to enhance performance, both physical and mental, and to facilitate pleasurable experiences—ecstasy, excitement, satiation, tranquility, or altered perception and consciousness. Physicians may be involved when drugs are used for these less tradi-

tional medical functions, but they often are not. Physicians prescribe sedatives and tranquilizers for the anxious and sleepless,
antibiotics for travelers, and hormones to block ovulation, but
rarely amphetamines for athletes and never heroin or LSD for
fun. Nevertheless, all of these drugs are often distributed and
used outside of the medical model with important ethical and
social consequences.

Problems and Complaints

In general the public appears to be relatively satisfied with
the medical control of drugs used in the treatment of the sick.
However, there are some complaints in addition to some problems that do not lead to complaint. First, drugs are expensive,
and at least some of that expense seems to relate to medical
control. The separation between prescriber and consumer interferes with market regulation of prices, and results in higher
prices, and as a result of those prices an increased risk that
drug prescriptions will never be filled. Second, there are major
financial and industrial interests linked to the manufacture and
sale of drugs, and this influences medical treatment in several
ways. It provides economic incentives for the discovery and development of some new drugs, although unfortunately this is
true whether or not the new drug is superior to older ones. It
fails to provide adequate incentive for the development of
other drugs, particularly those that cannot be patented or lead
to a profitable return. The lag in the development of lithium is
an example. Although this substance is widely regarded as one
of the most important drug therapies introduced into psychiatric practice in the past twenty years, it could not be patented,
had limited potential for profit, and therefore was not marketed with the same enthusiasm as many less important drugs.
The marketing efforts of the drug industry provide the most sophisticated and effective continuing education for many practicing physicians, but to a certain extent the goals of these educational programs relate more to selling products than to
optimizing patient care. In sum, many pharmacological treatments are developed more rapidly, marketed more effectively,
and utilized more widely than other treatments in medicine because of the interests of the pharmaceutical industry.

A third set of problems relates neither to the cost of drugs nor to the impact of the pharmaceutical industry on prescribing practices, but rather to the authority vested in physicians by the law. A physician's prescription is required to obtain many drugs, although a layman can purchase firearms or poison, risk his life in a variety of ways, or obtain nonpharmacologic therapy from quacks or charlatans without first obtaining medical permission. Some feel that this requirement for medical permission to obtain drugs is an inappropriate restriction of personal liberty.

Drug Use Outside Medical Control

Both the public and the medical profession have been much less satisfied with the use of drugs outside of the medical model. There are major public health problems related to the abuse of tobacco and alcohol, and continuing questions about the health consequences of marijuana. The use of cocaine, heroin, and related drugs is associated with widespread social, legal, and moral concerns. Although these concerns are often justified in terms of health, they seem to be based on other issues as well, and correlate only poorly with known deleterious effects on health. The nonhealth issues that contribute to negative or cautious social attitudes toward the use of a drug include: (1) recent invention or discovery; (2) popularity among low status groups in terms of ethnicity, socioeconomic status, or age; (3) tendency to be habituating or addictive; (4) rapid response that includes socially apparent alteration of mental state or behavior; (5) criminal involvement in supply or distribution; (6) administration by injection; (7) general belief that use of the drug is involved in a socially disapproved life style; (8) general belief that use is socially contagious; and (9) general belief that use is one step in a causal chain that leads to noxious social consequences, including the use of other more harmful drugs and criminal behavior. The most basic public health question, whether use of the drug causes disease or predisposes to further disease, particularly over a long period of time, is a relatively weak determinant of social concern, as evidenced by the general social acceptance of those nonmedical drugs that are associated with the most well-established negative effects on health—tobacco and alcohol.

When Is a Substance a Drug?

Some drugs, primarily sedatives and stimulants, are used both within and outside of the medical system, but most drugs used by doctors and patients are of little interest to other citizens, and, conversely, many of the drugs popular for pleasure and amusement have little value in the practice of medicine. This raises the question of whether the term "drug" relates to a single class of substances, or whether the substances prescribed by doctors and the substances purchased on street corners, in spite of a small overlap, represent fundamentally different categories and the whole issue of the medical model in reference to drugs used for pleasure or performance grows out of semantic confusion. Science offers us little assistance, for it is clear that the term "drug" is a social concept rather than a chemical or biological category. It is defined by the role that a substance plays in the transaction between a doctor and a patient, or a "pusher" and a user, and the purpose or intent in its use. One scientist suggested that the definition of "drug" was any substance that, when injected into an animal, produced a scientific paper, a recognition that even in the world of science, drugs are defined by their social consequences. Whether a given substance is a food or a drug is a matter of social decision and consensus, not of scientific fact. Alcohol has nutritional value, but we often regard it as a drug; coffee has no nutritional value and clear pharmacologic effects, but we generally regard it as a food; nutmeg may be food for some and drug for others.

What leads to the notion that a substance is not only a drug but, further, that its use should be controlled or regulated by doctors? First, it should be used by doctors in treating patients and not have any other widespread socially sanctioned role. Alcohol is used by doctors, but has many other socially approved uses, and, therefore, although a drug, is not a medically controlled one. Second, it must have some danger, risk, or at least some social cost. Aspirin is a drug used by physicians, but is not perceived as having sufficient risk to warrant medical control. Some drugs may have little risk, but such extraordinary cost that medical regulation is used to ensure their equitable distribution among those who need them most. Vaccines or antisera would be examples. Rare golden caviar may be equally expensive, but it has no increased value to the sick or vulnerable, and therefore medical regulation offers no advantages over the usual marketplace method of distributing scarce resources. The danger or risk that leads to medical regulation may relate to the community rather than the consumer, so that indiscriminate overutilization of antibiotics leads to the general danger of the emergence of resistant strains of microorganisms as well as health risks to individual patients.

The Medical Model: The Doctor

The social value of the physician and of the medical model in the regulation of drugs is based on the general social value of both the profession of medicine and the medical model. The term "medical model" has no simple or single meaning, but we shall take it to be a sociologic construct that describes the roles of physicians and patients, and the attitudes of others toward them. Contemporary medicine also has roots in scientific biology, but these are recent and provide only the current content of medicine. The profession is far older than the science, and doctors, patients, and drugs all antedate modern biology. A corollary is that the medical model has little to do with the organic etiology of various dysfunctions, and much more to do with illness and caretaker behaviors.

The doctor's role has three characteristics that are important in this model. First is expertise, the scientific and other knowledge and skill necessary to treat the sick. This comes from training and experience, and includes knowledge of pharmacology and materia medica, drugs and their uses, effects, and toxicity. Possessing such knowledge, the doctor is a good person to inform patients about drugs, to offer advice, and at times when information and advice are not sufficient, to make decisions relating to drug use.

A second characteristic of the physician in the medical model is his morality. Medical students are selected for moral as well as intellectual capacity; they are socialized into a profession that has an important moral tradition; and they are exposed to an unusual array of experiences that shape and develop moral sensitivity. As a result, doctors bring to decisions about drugs not only their knowledge, but also a set of values and traditions that shape their attitudes. One of these values is to place great emphasis on doing no harm, and related to this is the general reluctance of physicians to use drugs or other potentially dangerous treatments in the absence of disease.

Finally, the physician has authority in the eyes of patients. This has been called charismatic authority, a term that suggests that it stems from personal characteristics of the physician, but one does not have to look far to learn that physicians are personally extremely heterogeneous, and that many are far from charismatic. What unites them is not what they are, but rather how their patients perceive them. Their authority is based not on their charisma, but rather on the universal attitude of the sick or helpless toward the caretaker, the idealization that stems from the psychological relationship of the helpless child to the parent and is a latent potential in every adult reactivated in times of stress or anxiety. This phenomenon, labeled transference by psychoanalysts, is at the heart of the authority with which physicians are invested by their patients. None of these three characteristics is unique to the physician, but their combination is. Pharmacists have technical knowledge about drugs and pharmacology; priests have an ancient moral tradition; judges have authority that is largely based on transference, but each of these groups lacks at least one of the other essential characteristics of the physician. Their combination defines the profession of medicine and, as a result, the physician's role in the so-called "medical model."

The Medical Model: The Patient

In the clinical variant of the medical model, the role of the physician is coupled with the role of the sick patient. Parsons' classic formulation of this latter role lists four characteristics: (1) the exemption from responsibility; (2) the involuntary nature of the condition; (3) the desire to recover; and (4) seeking and cooperating with treatment.

It is clear that while all three of the characteristics of the physician—knowledge, morality, and authority—might be relevant to a potential role in the distribution of drugs for pleasure and performance, at least the first three of the four characteristics of the sick role do not apply. The fourth, seeking and cooperating with the prescription, would be relevant only if it were recognized that the prescription in this case had nothing to do with treatment.

A tentative conclusion is that while the physician's role might be of interest in the social regulation of drugs used for pleasure or enhancement, the patient role and therefore the clinical variant of the medical model is not.

The Public Health Model

There are, however, other nonclinical variants of the "medical model" that do not involve patients or the sick role. The most well known is the public health model, in which the physician's contract is with a community rather than an individual patient. The concern is with preventing or limiting the spread of disease as well as treating the sick, and the physician's actions frequently inconvenience or disadvantage some individuals in order to promote the health of others. Quarantines, vaccinations, and the treatment of asymptomatic disease carriers are all examples. This version of the medical model would seem to have some relevance to our problem, since many of the fears regarding the use of drugs for pleasure or performance relate to the impact of that use on the rest of the community. However, there is an important distinction to be recognized. The widespread concerns about the use of these drugs, as discussed above, are only distantly related to health, and are much more closely linked to general social and moral values. The physician is not expert in these areas, and previous excursions of physicians into general questions of the organization of society and the quality of life, such as in the community psychiatry movement of the late 1960s, have been largely discredited. In sum, the public health variant of the medical model is of relevance to our central question, but primarily as a metaphor.

The Medical Model: Disability

A third variant of the medical model involves a patient (or client), but as disabled rather than sick or diseased, and focuses on support and rehabilitation rather than cure. The disabled individual, like the sick one, is exempt from responsibility, is an involuntary occupant of the role, and is expected to seek and cooperate with appropriate professional help. However, by definition, the disabled individual cannot be "cured," and the goal is optimal adaptation, with the assumption that deficits will remain. The physician's role in this model often involves attempts to enhance existing normal functions rather than alter diseased ones—as in the rehabilitative programs for neuromuscular disease that emphasize the development of compensatory motor capacity or the enhancement of alternative channels of communication in the blind, deaf, or aphasic. Few would question the use of a performance-enhancing drug that would improve normal capacity to above normal and thus allow a disabled individual to lead a more regular life, as in those individuals with severe spasticity who find that they can control their limbs more easily with the muscle relaxing effects of "minor tranquilizers," or the use of drugs that produce euphoria while relieving discomfort in those terminally ill patients who are disabled by chronic pain. In fact, one of the most distressing side effects of the social concern with the use of narcotic drugs for pleasure has been the extension of the general moral disapproval of these drugs in such a manner as to contaminate medical judgment, with the result that physicians now tend to underprescribe narcotics for patients with severe pain, as though there were something inherently evil about drugs that produce pleasure regardless of the setting or purpose of their use.

The Medical Model: Enhancement

A fourth variant of the medical model may provide the closest analogy to the use of drugs for pleasure or the enhancement of performance. It is a clinical model, but the patient is neither sick nor disabled. It is rather that the technology or treatment involved stems from medicine, or involves temporary risks or disabilities that lead to a time-limited voluntary assumption of the sick role. The medical profession tends to be uneasy with this variant and its potential abuse, in much the same way that society is uncomfortable with the nonmedical use of drugs. There are controversies regarding whether the costs associated with these "treatments" should be regarded as expenses for health, and whether they should be reimbursed by medical insurance. In general, the treatments involve attempts to enhance the pleasure or performance of individuals who are socially regarded as "normal" or "healthy," but who have greater potential. Examples would include some aspect of cosmetic surgery, sports medicine, sex therapy, psychoanalysis, and treatment for insomnia, premenstrual tension, or the symptoms of menopause. The "patients" involved are not sick in the usual sense, but medical treatments are widely employed and doctors administer those treatments.

What Are the Options?

Where does this leave us? Doctors have some characteristics that are relevant to the regulation and distribution of drugs, but their experience and expertise in prescribing drugs stems primarily from their role in the doctor–patient relationship and the first clinical variant of the medical model. The use of drugs for pleasure or the enhancement of performance lies outside of this most familiar medical model, although there have been some extensions of medicine into the supervision of medically related technology for patients who are not sick in the usual sense, the fourth variant. The public health role of physicians is also relevant, but only insofar as the concern about these drugs relates to their impact on health, and particularly to their impact on risks to the health of the community rather than the user. At present, this would seem to be a limited cause for concern. Finally, doctors currently prescribe drugs for both pleasure and the enhancement of performance, but they generally do so only for the disabled, as in the third variant of the medical model, with the hope of minimizing disability by enhancing some normal functions to above average in order to compensate for others that are defective. This model is an interesting one, for it directs attention to one of the general social concerns about the use of drugs outside of the first clinical medical model; namely, that certain aspects of the disabled role are inherent in the use of drugs (or other medical treatments) in the absence of disease. The fear is that the psychology of the drug user will assume aspects of the psychology of the disabled. It may be this as much as physiological dependency or addiction that is at the core of the social phobia of drugs. Indeed some of the literary fantasies of future societies that rely on drugs for pleasure suggest citizens whose personalities are reminiscent of pathological adaptations to disability.

The options seem clear. First, we might choose to make drugs freely available outside of the medical model. The obvious advantages include an enhancement of individual liberty, encouraging the scientific development and evaluation of useful new drugs, and perhaps a decrease in their cost. The disadvantages include tinkering with a social institution that has functioned moderately well for a long time, shifting control of a potentially dangerous technology from a group who generally have the required expertise (physicians) to a group who generally do not (the public), and removing a social barrier to an activity that many fear might threaten some core aspects of the psychological sense of self and autonomy and therefore endanger central social values.

Alternatively, we might continue to restrict the control of drugs to physicians, while at the same time allowing them to determine the somewhat vague boundaries of disease and patienthood. This system has many "leaks," as drugs are now widely used outside of medical regulation. It has major inequities, as prominent athletes or public figures are widely known to obtain drugs from medical sources that are not available to the general public (although it is sometimes difficult to decide who is being discriminated against when bad treatment is selectively made available to an elite). It also involves a fundamental decision to continue to define drugs in a traditional context, as a mysterious and potentially dangerous or evil substance beyond the understanding of the common man, and best restricted to the control of a select few. However, this definition has become inconsistent with the increasing specificity of drug effect, growing scientific understanding of drug action, and wider public education about drug use.

There are also a number of intermediate models possible. Drugs can be classified as strong and weak, safe and dangerous, traditional and experimental, or medical and nonmedical. In effect, current policy exempts weak, safe, traditional drugs from medical control, while regulating strong, dangerous, experimental ones, and making judgment calls in mixed cases.

Conclusion

Whether or not individuals should use drugs for pleasure or the enhancement of performance is an important social and ethical question. Physicians have some relevant expertise, particularly concerning the pharmacology of the drugs, the possible medical risks, and the psychological correlates of drug use. They are not experts on the basic social or moral questions, and the medical profession is not the appropriate locus of decision-making concerning them. The medical model is peripherally applicable to the control and distribution of these drugs, and is available as a means of such control for those situations in which society desires regulation. It is important that such social regulation as does occur does not impair the legitimate medical utilization of such drugs for the sick or disabled, as already has occurred in the use of narcotics. Finally, many of the problems concerning the use of drugs stem from the lack of public education, and continued efforts to disseminate information about drugs, their actions, effects, and usage are desirable.

Drugs, Models, and Moral Principles

Ruth Macklin

Introduction

The complexity of the ethical, social, political, and medical issues surrounding the use of drugs and attempts to arrive at sound public policy needs no reminder. The articles in this volume serve to underscore some things we already know, and to deepen our understanding both of the complexity of the issues and of the forces that have led to abysmal failure in the many and varied attempts to deal with our society's "drug problem." It may surprise some readers to learn that a special form of "drug problem" exists in the world of sports—both in Olympic competitions and professional athletics—a problem stemming from the use of drugs both as performance enhancers and as "performance enablers," as John Conrad puts it. The use of cocaine and other performance enhancers falls outside the medical model, whereas the administration of painkillers to injured athletes falls squarely within that model. Moral principles apply to both of these uses of drugs by athletes, but they are different moral principles in each instance.

It may also surprise some readers to learn that social attitudes toward risk-taking behavior are somewhat inconsistent in our society, leading to considerable confusion over the justification that can be offered for government intrusion into the risk-taking behavior of drug users. Bakalar and Grinspoon offer some explanations for these attitudes and purported justifications, and their essay illuminates this topic by showing the different models at work governing social attitudes toward drugs and the attendant social policies. Government prohibition of risk-taking behavior by citizens may at first appear to be a straightforward example of state-enforced paternalism, or "legal moralism" as it has been called. But the essays by Bakalar and Grinspoon, Neville, Michels, and Schwartz demonstrate the mistaken simplicity of an account that seeks to characterize control of drug users solely in terms of opposition between the values embedded in paternalism, on the one hand, and individual liberty, on the other.

Perhaps the most striking fact that emerges from the observations and reflections in these essays is the sheer number and variety of moral principles and value considerations at play. The aim of this concluding chapter is to identify and elucidate those principles and values, to discover which ones are in tension with one another, and to explore when disagreement over drug policy stems from adherence to different moral principles and when it arises out of different assessments of the facts. This inquiry will, alas, not yield a definitive normative conclusion about the propriety of individuals' use of pleasure or performance-enhancing drugs. However, it should serve to provide a deeper understanding of why problems regarding drug use in our society are so intractable. To grasp that point clearly, we need to recall that a surprising number of moral controversies stem from disagreements about the facts of the matter, while others typically result from clashes of ethical principles. Both sorts of disagreement are present in the drug controversy, and it is their inextricable entanglement that makes this problem so hard to understand clearly, much less to resolve.

A final complicating feature is that of the different models used to conceptualize the issues and to formulate policy regarding drugs. At least three so-called models have been invoked: the medical model, the moral model, and the enforcement model. The medical model itself has several variants, as Robert Michels shows, and the fact that mixed models have been operative in the social and political sphere relating to drugs makes it all the more difficult to gain a clear understanding of the problem as well as to fashion a resolution.

Moral Principles

Let us begin, then, by listing the moral principles and related value considerations that play a role in controversies over the propriety of drug use and attempts at regulation. There are many ways of characterizing and classifying ethical principles, and the schema presented here has no unique claim to correctness. I propose this way of looking at these complex factors to assist in grasping the many threads woven through the preceding essays.

The moral principles and other value considerations invoked in the foregoing chapters can be classified under three broad headings: (A) *consequentialist* principles; (B) *nonconsequentialist* principles; moral and legal principles based on *rights*; and (C) precepts of *intrinsic* value. The breadth of these general categories is such that the principles falling under them include: the purely moral; those that are at the same time moral and legal; constitutionally derived legal principles; and political principles. Although some of the ethical notions invoked in connection with drug use by individuals and drug policy formulated by the state are clearly recognizable under these broad headings, others require further elucidation. The task is to show just what values are embedded in them, and to see whether those values appear in the form of individual rights and duties, the health, happiness, or well being of persons, the legitimate interest and authority of the state, or the solidarity, fabric, or welfare of society as a whole. That the topic of nonmedical uses of drugs gives rise to all these types of moral principles, and the inevitable conflicts between and among them, suggests this as a rich case study for ethical, legal, and political analysis.

Consequentialist Principles

Consequentialist moral principles make the value of an action, practice, or social institution contingent upon its actual or probable results or consequences. Deciding on the rightness or wrongness of an act or practice is a matter of looking at the results or outcomes, compared with those of other possible acts or practices, and determining which act or practice produces a balance of favorable consequences over unfavorable ones. Of course, consequentialist theories of ethics must provide a criterion for the goodness or badness of outcomes or results: what characteristic of the consequences is to count as the good-making or bad-making property? Loosely construed, that general quality is usually characterized as happiness or unhappiness, pleasure or pain, welfare or illfare, with these terms referring either to the sum of individuals affected by the action or to society as a whole.

The best-known consequentialist theory of ethics is utilitarianism. In its classic form, the principle of utility holds that right actions are those that produce a balance of pleasure or happiness over pain or unhappiness for all those who stand to be affected. John Stuart Mill, the 19th-Century British philosopher who was one of the staunchest proponents of utilitarianism, referred to this as "the greatest happiness principle." Applications of the utilitarian principle appear under other names as risk–benefit and cost–benefit analysis. The use of cost–benefit analysis as a tool of economic decision-making rests squarely on the utilitarian principle, as does the more general precept of risk–benefit, dictating choices that minimize risks of harm and maximize probable benefits.

The successful application of any consequentialist principle depends crucially on an accurate assessment of the facts. Thus the theory is only as good as the scientific or factual evidence at hand, enabling predictions about the probable effects of practices or policies on individuals or on society as a whole. Although many claims regarding drug use and drug policy rest on alleged facts about the consequences of individual drug use and on the effects of certain social and legal policies, those claims are disputed by their opponents largely on empirical (as contrasted with ethical) grounds. Since consequentialist moral arguments rest on nonmoral, factual premises, the truth or accuracy of those factual premises is the key determinant of the validity of the argument.

Consequentialist arguments provide central themes in the papers by Bakalar and Grinspoon, Conrad, and Neville; they also appear prominently in the chapters by Zinberg, Brock, and Murray.

Bakalar and Grinspoon begin their account by contrasting drug use with other forms of risk-taking behavior considered socially and legally acceptable in our society. Except for alcohol, tobacco, and caffeine, drugs for pleasure or for the improvement of work performance are considered unacceptable.

> *This is certainly not because the dangers of drugs have been carefully compared with other kinds of danger, and not even because the risks of legal drug use have been carefully balanced against the benefits, or against the cost of enforcing punitive*

> *laws. Ironically, that if we did those things, alcohol and tobacco*
> *might be the first drugs banned. (p. 13).*

Later in their paper, the authors draw some parallels between prohibition of alcohol in the 1920s and current prohibitionist policies regarding marijuana. They note the same "nasty side effects" of these prohibitions: costs of arrest and punishment, disrespect for law, organized criminal violence, police corruption and oppression, poisonous adulteration, and misrepresentation (p. 25). Although Bakalar and Grinspoon do not draw the explicit conclusion that the costs of a prohibitionist policy at least toward marijuana (to consider only one banned drug) outweigh the benefits, they invite us to consider the reasons why such a cost–benefit analysis is unthinkable in our society, in stark contrast to the use of this decision-making tool in myriad other policy contexts.

John Conrad, on the other hand, devotes much of his essay to a direct appraisal of the cost–benefit ratio. Arguing against the absolute prohibition of nontherapeutic drugs, and in favor of some form of regulation of drug traffic, Conrad concludes that "although the regulation of a licit traffic in narcotics will be subject to abuses, it will produce benefits that the absolute prohibition of that traffic cannot bring about, and it will also remove some of the most destructive consequences of the present laws (p. 50)." A bit later he states:

> *In spite of enormous expenditures of public funds . . . , the*
> *diversion of police attention to the enforcement of these laws, and*
> *the investment of some remarkable tactical ingenuity and*
> *technological innovation, it has been a losing battle with heroin*
> *and cocaine and tacit surrender to marijuana. The results are dis-*
> *couraging, but the nation perseveres in the struggle to enforce*
> *these virtually unenforceable laws (p. 55).*

Much of Conrad's chapter is devoted to an effort to mount this consequentialist argument. Whatever other considerations may be thought relevant to formulating drug policy are subordinated to this predominant, utilitarian value. Conrad thus concludes by noting that the narcotics problem "will not disappear, it cannot be prevented, and the only hope for minimizing whatever harm it can do—or maximizing its benefits—must come from reasoned regulation" (p. 63). There could be no more explicit statement and acceptance of the utilitarian principle as the relevant, governing moral principle in this matter.

The chapter by Norman Zinberg continues the central themes in the essays by Bakalar and Grinspoon and by Conrad, but with a focus informed by Zinberg's own research on the importance of set and setting, in addition to pharmacology, for understanding the phenomena of drug use. In the details of his account, as well as in the policy recommendations at the end, Zinberg rests his analysis on a consequentialist footing. Agreeing with Conrad's main point, Zinberg states: "By now it has become clear that our present prohibition policy is being maintained at high cost" (p. 37). Among those costs are ones noted in a 1978 report from The President's Commission on Mental Health:

> . . . *loss of respect for the law in general since the drug laws were being persistently flouted; increased corruption among enforcement and other public officials; and a virtual consensus among informed citizens that although they might support the drug laws in principle, they would try to circumvent them if relatives or friends were involved (p. 40).*

Zinberg's own research, along with that of others, has shown that contrary to popular myth, continued moderate use of even presumably highly addictive narcotics is possible. Zinberg and his coworkers have also demonstrated the importance of informal social controls in influencing the perception, drug tolerance, and drug-related behavior or users. Why, then, does Zinberg come out in the end against decriminalization or legalization of all illicit drugs? The answer is again a consequentialist one.

Zinberg cites as a typical example of the *laissez faire* stance the views of Thomas Szasz. Szasz serves to illustrate not only the adherent of the libertarian political principle in its most extreme form; he also exemplifies the ideologue who accepts his political and moral precepts without argument and without the need for reasoned justification. In the domain of drug policy, as in the other areas in which he has written, Szasz is a staunch proponent of libertarianism, "whatever the consequences." Zinberg, in contrast, though recognizing the social, economic, and moral costs of our present drug policy, contemplates the consequences of abandoning formal social controls altogether. Citing the work of other scholars, Zinberg observes that "the increase in the number of drug users that would result from such an approach would mean an increase in at least the absolute number of drug casualties" (p. 43). The goal of social policy, according to Zinberg, should be to regulate use and to prevent abuse. Close attention to the consequences both of drug use and of public policy is reflected in Zinberg's specific recommendations at the end of his chapter. For example, he strongly encourages making a distinction between the two major types of drug use: moderate occasional use, which incurs only minimal social costs, and the more dysfunctional and compulsive use patterns. Zinberg's conclusion represents a balanced consequentialist judgment, flowing from an honest appraisal of the drug problem and our society's attempts to grapple with it: "Thus, other necessary ingredients in our effort to avoid the serious consequences of intoxicant use, such as treatment programs, legal reform, and medical research. . .would lead to further cautious steps to reform our current policies and practices" (p. 44).

Turning next to the essay by Robert Neville, we find at once a combination of consequentialist and nonconsequentialist moral principles. Neville terms his two leading principles "abstract political principles"; they are "political" because they address the question of state intervention into individual behavior. Since Principle II, the consequentialist principle, is a qualification of Principle I, both will be discussed at this point.

Principle I, in the form Neville states it, is a nonconsequentialist principle of private autonomy: "With proper qualifications there is no justification in the state's intervening in direct, responsible transactions regarding drugs between users and dispensers, or in the possession and use of drugs by individuals" (p. 66). Principle I makes no reference to consequences or outcomes of the policy it states. Instead, it rests on the underlying premise of the political tradition of liberalism: except for circumstances that may bring harm to others, individuals should be left alone in their own pursuits and transactions. Put another way, for the state to intervene in the behavior of its members, some justification is needed that would permit overriding the sovereignty of individuals in matters concerning only themselves.

Neville's Principle II states the conditions for justified intervention by the state: "The public consisting of those indirectly involved in transactions regarding drugs, their possession and use, by virtue of suffering the consequences of the transactions, possession and use, has a justified interest in intervening to control those consequences" (p. 66). Unlike a very different political principle, one that embodies legal moralism by permitting the state to intervene in the behavior of its members for their own good, happiness, health, or welfare, this principle justifies intervention only to the extent that bad consequences accrue to individuals other than the drug users themselves.

Neville is careful in the way he constructs his consequentialist argument. Having stated that people are justified in participating in any social process that affects them, he concludes that if individuals' drug behavior has consequences for other people, that is an invitation to the other people to intervene in the interests of controlling those consequences (p. 66). It does not automatically follow that it is the state that has proper authority to intervene; rather, Neville maintains, an extra argument is required to move from that narrowly defined public (those suffering the consequences of drug use) to the state as the agent of intervention (p. 67). The extra argument appears toward the end of the chapter, where Neville offers two reasons for holding the government to be the proper intervening agency. First, "the scale of the economic base of drug use makes the entire citizenry the relevant public as represented in government"; and second, "government is the relevant agency for handling consequences of behavior that are distributed according to such anonymous or public forms of social interaction" (p. 77). Like Bakalar and Grinspoon, and Conrad, Neville remains alert to the consequences of intervention itself: "the personal and social costs of intervention are very high and therefore are among the most important values to be considered in weighing what to do" (p. 77).

It is notable that Neville does not ground his political principles in a theory of rights, whether rights of individuals or of the state. Although this may not seem of central importance to those concerned more with the practical implications of public policy than with its theoretical roots, it nevertheless remains crucial for an understanding of the derivation of moral principles widely invoked in policy contexts. Neville states that his own view departs from classical liberalism in making "rights secondary to participation rather than the other typically liberal way around" (p. 67); most liberal and libertarian accounts begin by asserting the rights of individuals against the state. The core concept in Neville's analysis is that of participation in the social process, a more communitarian notion than the precept of individualism found in liberal political theory.

A final consideration emerges in Neville's paper, a consideration focusing not simply on consequences of drug use for users themselves or others in their immediate environment, but on larger social consequences. Since this same consideration is also raised in several other essays, it would be well to treat it as a special subprinciple under the more general consequentialist principle discussed so far. Bakalar and Grinspoon refer to this possible range of consequences as "a threat to the social fabric and the moral order" (p. 18). They refer also to "community bonds" and to more precise phenomena, such as "productivity" and "public tranquility" (p. 20). Productivity is also mentioned specifically by Neville in his discussion of the texture of society (p. 72), and Robert Schwartz invokes notions such as the social need to "maintain the fabric of society" (p. 146) as well as public tranquility, the element noted also by Bakalar and Grinspoon. Let us call this, for convenience, the "fabric of society" argument, treating it as a version of consequentialism that looks to alterations in society as a whole, rather than to effects on discrete, identifiable individuals.

Conducting a thought experiment, Neville invites us to consider the development of a genuine Soma drug, one that would have no harmful effects on the user. Neville asks how use of this Soma would alter individuals' lives, and quite correctly observes that the answers to these questions can only be given by "serious social study." Neville raises these issues as part of his inquiry into what it means to use drugs "responsibly," but we may consider this consequentialist argument from another perspective. An argument that looks to the potential damage to the fabric of society is distinct from several others with which it might be confused. It is not the same as a paternalistic argument, since it is not the protection of people from their own self-destructive behavior that is the aim of social policy designed to preserve the fabric of society. Nor is it a version of the "harm principle," a form of Neville's Principle II justifying interventions into some people's behavior in order to prevent harm to other (innocent) persons. Since it is not harm to the user of drugs or specific harms to identifiable others that is at stake, what sorts of consequences are these? Schwartz refers at several points to "esthetic" considerations (pp. 146), but he does not mean to limit these concerns to those that might be deemed "merely" esthetic. They seem to fall best under the heading of "quality of life" consequences—somewhere in between full-scale moral consequences, such as harm to persons or direct harm to the interests of individuals, and esthetic consequences. Whatever we mean by "the fabric of society" or the "moral order," it is something more serious than litter or graffiti (important esthetic concerns, to be sure) and something less serious than anarchy, terrorism, or social mayhem. This rather nebulous social consequence is worth keeping in mind as a distinctly different matter from the purported consequences to drug users themselves and to their unwitting victims, as those groups are portrayed in popular mythology and in so-called drug education courses in their typical secondary school form. If the "reefer madness" myths were exploded, once and for all, by careful research and a balanced program of public information, it would leave untouched the separate argument concerning the fabric of society.

Of the two remaining chapters that make explicit use of consequentialist moral principles, each addresses the avowed purpose for which drugs are ingested. Dan Brock explores the use of drugs for pleasure, whereas Thomas Murray examines drugs as performance enhancers in the context of sports. In both essays, consequentialist principles are not the only ones that enter the argument, but the discussion here will be confined to those principles.

Much of Brock's essay is devoted to an elucidation of philosophical theories of pleasure and to an examination of the good for persons. None of these general philosophical accounts of pleasure yields any specific conclusions regarding the propriety of drug use for pleasure, at least in the absence of a considerable body of empirical data. These data mostly concern the effects of drug use, in particular, effects on the user. Tracing the existence of disputes to this factual matter, Brock notes:

> I believe that most of the disagreement concerning the desirability of the use of drugs for pleasure rests not on philosophical, and more specifically moral, differences concerning the good for persons, but rather on disputes over the empirical facts concerning the consequences of the use of drugs for pleasure (p. 101).

The empirical information needed is of two sorts: first, information about persons and what makes them happy; and second, information about the various effects that use of drugs for pleasure has. Providing a sketch of what most people would agree are the basic components of the good for persons, Brock then invokes the popular image of the drug user, suggesting that the features of this image do indeed prevent the realization of the specified goods. Yet he reminds us again that this is a matter to be determined by empirical studies, not by philosophical analysis. Brock concludes by stating his belief that

> . . . it is on this issue of the nature of the consequences of the use of pleasure-producing drugs that rational controversy concerning the evaluation of their use must largely turn, and not on philosophical and evaluative differences about the intrinsic value or disvalue of the use of drugs for pleasure (p. 106).

Having examined the main philosophical theories of intrinsic goodness, Brock concludes that they have no clear implications for the propriety of drug use for pleasure. Once again, we must look to the consequences as a determinant of the rightness or wrongness of this type of behavior.

Thomas Murray explores several objections resting on ethical grounds to the use of performance enhancing drugs by athletes. He rejects as unsatisfactory arguments that rely on duties to oneself, and dismisses also a more complex argument to the effect that by harming ourselves we may violate our moral duties toward others. The argument Murray fastens on is one that is fundamentally consequentialist, but it may also be construed as an argument from justice or fairness. The argument asserts that "given the social nature of the enterprise, performance-enhancing drug use in sport is inherently coercive" (p. 119). The argument rests, first of all, on facts about sport as a social institution: athletic events are intensely competitive, hence there is tremendous pressure on players, coaches, team physicians, managers, owners, in short, everyone connected with a team, to seek a competitive advantage. Given these pressures, athletes are coerced into the same patterns of behavior as their opponents and their own teammates, and so lack a genuinely free choice in the matter. The option of seeking competitive advantage is what William James called a "forced choice"; in Murray's terms, athletes are pressured into using performance-enhancing drugs, or leaving the competition.

The argument from inherent coerciveness is, in large measure, a consequentialist one. Murray constructs the argument as follows:

*My alleged liberty to take performance-enhancing drugs,
which is very hard to counter from an individualistic conception
of morality, is counter-balanced by the pressure I place on my fellow competitors. My "free" choice contains an element of coercion. If enough people like me choose to use performance-
enhancing drugs, then the freedom of others not to use them is
greatly diminished (p. 123).*

This argument is a complex one, since it requires an assertion
of the value of individual liberty while at the same time
acknowledging the limits of a principle of individual liberty
when taken alone. Murray's argument points to the consequences for the liberty of others when some athletes use drugs
to enhance their own performance. Hence it is a mixed argument, employing as premises both a principle of liberty (or
noncoercion) and also a factual premise about the effects drug
use by some players has on other players. Murray's argument
would not go through without either premise: the noncoercion
principle is needed to show that drug use by athletes is morally
wrong; and the empirical premise is needed to show that freedom on the part of some athletes to use performance-
enhancing drugs does, in fact, have the consequence of
coercing other players (limiting their liberty) into doing the
same.

There is, finally, in Murray's argument a trace of the harm
principle, as well. In reviewing his conclusion, Murray writes:
"despite the claims of individual autonomy, the use of
performance-enhancing drugs is ethically undesirable because
it is coercive, has significant potential for harm, and advances
no social value" (p. 125). We must not overlook the "potential
for harm," since it is chiefly on this ground that some athletes
(especially women) seek to avoid the use of steroids and other
hormones that have undesirable side effects in addition to the
desired effects of enhancing strength, stamina, and other elements providing competitive advantage. Whether they pose
genuine health hazards, interfere with reproductive capacity,
or induce esthetic or other socially undesirable changes in ap-
pearance, the harm caused to the user by steroids counts as a
morally justifiable reason to remain free of coercion. Not all
forms of coercion are automatically unjustified; instances of coercion may be justified where some overwhelming good is
likely to be achieved by the coercive measures. Although winning at sports, be they amateur or professional contests, is

viewed by many as an overwhelming good, a more balanced judgment must look to the costs of health hazards to athletes coerced into using performance-enhancing drugs. It is, then, the consequences of coercion itself, coupled with the prospects for harm to drug users, that clinches the ethical argument Murray offers against the use of performance-enhancing drugs in sport.

I said earlier that although Murray's argument is fundamentally consequentialist, it may also be construed as an argument from justice or fairness. Now moral principles that relate to questions of justice or fairness are generally nonconsequentialist in nature, so it may be puzzling how justice enters in at all here. To state my point more precisely, the ethical considerations raised by the use of performance enhancers in sports include issues of justice. But the way Murray constructs his argument, and the salient social facts he incorporates among his premises, preclude the need for principles of justice. The point is this. If some athletes used performance enhancers in their competitions, while others did not, those who did would have an unfair advantage. The argument against drug use in that case would rest on a principle of justice as fairness. Although drug use might, in principle, be open to all (albeit illegally), those who refused to use performance enhancers, for whatever reason, would be at an unfair disadvantage. Since the concept of sport entails some notion of fair competition among the contenders (at least ideally), any departure from those precepts of fairness would be unethical.

That is a perfectly good argument, where it applies. The fact that Murray did not employ that argument, or even invoke the concept of fairness (with one small exception, while making a different point on p. 117), demonstrates his confidence in the truth of the factual premise of his own argument, to wit: "Olympic and professional sport, as a social institution, is an intensely competitive endeavor, and there is tremendous pressure to seek a competitive advantage" (p. 122). Conjoined with the well-documented facts about the availability of drugs to athletes and the prevalence of the use of a wide variety of performance enhancers and performance enablers (pain killers), Murray could construct his ethical argument in terms of inherent coerciveness rather than in terms of justice or fairness. But the latter concepts remain relevant to the general issue of the use of performance enhancers in competitive endeavors,

whether those endeavors are sports or examinations, and
whether the performance enhancers are chemicals or other aids
to performance that place some people at an unfair advantage
vis-à-vis others in the competition.

Consequentialist principles regarding drug use and regulation are by far the most numerous of the several varieties that
appear in these chapters. Some of the other candidates for ethical principles applicable in this domain were mentioned in
passing here, and they will be discussed briefly in what follows. But it is worth noting at this point that conflicts of principle are one of the earmarks of moral dilemmas. One of the
sources of ongoing controversy in numerous areas of public
policy is the conflict between relevant consequentialist principles (in particular, cost–benefit analysis) and principles that
invoke individual liberty or freedom, among other values. It
would not be surprising to find those conflicts embedded in
controversies surrounding the "drug problem" as well.

Nonconsequentialist Principles; Moral and Legal Principles Based on Rights

By far the most well-known and widely invoked
nonconsequentialist moral principle is some version of the classical liberal principle of freedom or liberty. Libertarian moral
and political theories couch this principle in terms of the right
of individuals not to be interfered with for any reason except to
prevent harm to others. Even a nonlibertarian theory such as
the one as John Rawls sets out in *A Theory of Justice* contains as
its first principle a "principle of equal liberty" couched in terms
of rights: "Each person is to have an equal right to the most extensive total system of equal basic liberties compatible with a
similar system of liberty for all."[1] Given the ascendancy of
rights talk in the moral and political spheres in recent decades,
and given the preeminence of the liberty principle in our country's heritage, it is striking how little explicit mention there is in
the foregoing chapters of libertarian rights. To be sure, a number of essays mention the precept of liberty or freedom, but
these claims do not appear in the language of rights, and they
are not tied to a clearly articulated political or moral theory.

Several places where the value of individual freedom or
liberty is mentioned are as follows. Bakalar and Grinspoon re-

mark: "Permitting drug use is sometimes defended in the name of individual freedom, but rarely on the ground that there is any good in it" (p. 15). Conrad observes, first:

> *If adult citizens have had the opportunity to inform them-selves of the consequences of drug use and choose to take whatever risks there may be, it is not possible to reconcile the state's inter-ference with the principle that citizens should be allowed freedom to the extent that their actions do not infringe on the rights of others (p. 61).*

And then: "Adults should be free to decide for themselves whether or not to use narcotic drugs" (p. 62). Robert Michels, discussing the medical model and physicians' control over prescribing drugs, states: "Some feel that this requirement for medical permission to obtain drugs is an inappropriate restric-tion of personal liberty" (p. 177). None of these authors em-ploys the language of rights, and none makes the attempt to provide a sustained argument in support of their own or oth-ers' contentions about individual liberty to use drugs.

As we saw earlier, Robert Neville's Principle I is a version of the classical liberal principle, but Neville terms it a "principle of private autonomy" rather than one of liberty or freedom. Furthermore, Neville notes explicitly that the notion of rights to which he subscribes is one derived from participation, "rather than the other typically liberal way around." His Principle I makes no mention of rights, or of the liberty or freedom of indi-viduals. As we also saw earlier, Thomas Murray devotes some time to the question of individual liberty in the context of sports in which athletes use performance-enhancing drugs. But Murray also avoids couching his own conclusion or the views of others in terms of rights. Murray argues that it is wrong to interfere with the liberty of others, but he nowhere states that individuals have a right not to be interfered with.

It is refreshing, indeed, to find so little use of rights talk (which is usually introduced for rhetorical force) in the prece-ding chapters. The overuse of rights language has diminished its significance in the situations in moral and political life in which genuine rights are being violated. The two essays in this volume devoted at length to questions of rights are by legal scholars, each exploring an area of constitutionally mandated rights and their applicability to the use of drugs. The rights un-der discussion are the right of privacy, addressed at length by

Robert Schwartz and also by Nancy Rhoden; and the First Amendment right to freedom of thought, addressed by Rhoden.

Rhoden undertakes an extensive comparison between the right to refuse psychoactive drugs, a right granted in a number of recent court cases, and the right to use such drugs, based on analogous constitutional precepts. Schwartz probes a number of other areas of law in which rights of privacy have been upheld, and examines their relevance and significance for the right of individuals to use mind-altering drugs. Both chapters rest their analyses on firm legal grounding, primarily in the area of constitutional law. Rhoden concludes that the right to refuse treatment cases do tend to bolster arguments for a right to use mind-altering drugs, but the right in question appears to be more one of autonomy than privacy. She writes:

> *When we keep the comparison between the two classes of persons—mental patients and presumably rational adults who wish to use marijuana—firmly in mind, the right-to-refuse cases seem to lend a great deal of support to proponents of decriminalization of marijuana and other relatively harmless recreational drugs. In the right-to-refuse cases, courts are upholding the rights of persons seriously ill with mental and emotional disorders to refuse the very medication which could alleviate their emotional distress. Needless to say, at least some mental patients will be harmed by their refusal. It is, of course, possible that harm, either physical or psychological, will inure to some users of marijuana. However, such users will have chosen to take this risk, and there is little reason to believe their judgment is less rational than that of mental patients who refuse medication (p. 168).*

Schwartz's conclusion is more tentative. He holds that "the development of the Constitutional right of privacy does not dictate a model for federal and state prohibition and regulation of drug use. . . . The Constitutional analysis does instruct us that any regulatory scheme ought to be the consequence of careful evaluations of precisely what kinds of drugs are to be regulated for what purposes, among what populations, and under what circumstances (p. 150)." Although Schwartz's analysis is confined largely to that of one Constitutional right, the right of privacy, it is instructive in its thorough exploration of legal analogies that bear some resemblance to the moral issue of individuals' use of recreational drugs in our society.

The discussion of rights-based moral principles thus far has proceeded on the assumption that the rights in question all belong to the individual, as traditional liberal theory holds. However, even within liberal political thought (in striking contrast to *libertarian* political views), the *parens patriae* power of the state has been invoked in support of government paternalism. Sometimes couched in terms of the state's right to intervene in the behavior of its members for the sake of their health, well-being, or other interests, instances include the justifiability of civil commitment to mental institutions, the prohibition of substances (other than psychoactive drugs) believed to cause harm, such as laetrile and saccharin, the requirement that motorcycle riders wear helmets, court-ordered medical treatment for competent adults who have refused to grant consent, and other areas in which the law regulates the conduct of rational adults for their own good.

It is not surprising to find relatively little sustained discussion in this book of the justifiability of state paternalism regarding drug use, since the essays do not dwell at length on the countervailing moral/political viewpoint—the libertarian principle. The discussion that does appear focuses largely on the notion of consumer safety legislation, in the essay by Bakalar and Grinspoon, with an interesting twist added by Conrad. Since none of the contributors to this volume embraces the simplistic thesis that the ethical issues involved here can be characterized as a straightforward conflict between individual liberty and state paternalism, it is understandable that so little attention is paid to questions of justified or unjustified paternalism regarding drug use.

Robert Neville poses a direct question about justified paternalism in this domain, and offers a relativistic answer. Considering the case in which drug use is known to be harmful to the user (but not to others), Neville asks: "Does a principle of paternalism apply that would legitimate intervention to alter or stop the drug use?" (p. 70). He replies that the answer is relative to different social groups, depending on "the degree of organic tightness in their sense of social solidarity" (p. 69). Although that concept requires further elucidation, Neville makes it clear that our society is not one of tight solidarity, a conclusion with which few would disagree. As a result, Neville holds that "the imposition of paternalistic values would have to be imperialistic except in accidental cases" (p. 69).

Bakalar and Grinspoon examine the practice of drug control as an instance of consumer safety legislation. They use carefully selected analogies, such as motorcycle parts with a defective design or electric chainsaws, to show why many would argue that given the belief that marijuana smoking has no value of its own, the government has the right to protect people from it. "Drugs with no accepted medical uses are treated like the defective motorcycles; they are pleasure vehicles that have a deadly flaw" (p. 15). Of course, in order for any analogy to hold, the items compared must be similar in relevant respects. Bakalar and Grinspoon cast considerable doubt on the similarity in relevant respects between the hazards of chainsaws and defective motorcycle parts, on the one hand, and marijuana smoking, on the other. To the extent that such analogies break down, so too do the arguments designed to justify government intervention on the model of consumer safety legislation.

John Conrad offers a novel twist to the traditional paternalistic arguments regarding government intervention in drug use by individuals. Having already adopted a largely libertarian stance, premised on the demonstrable ineffectiveness of past and current drug policies, Conrad adds a new argument that rests on consumer protection. He notes that "what we have now is an unregulated market in which consumers use what they can get without guidance except by hearsay and 'street pharmacology' about the precautions to be taken" (p. 62). People are cheated and expoited by dishonest dealers, they take risks with adulterated or substituted substances, and there is no recourse through the usual channels that attempt to redress consumer fraud. Conrad calls for regulation by the government of substances now distributed by a criminal underworld. By making unstandardized drugs subject to government standards, the safety of the consumer can be protected. The argument, in effect, is that if the government were really interested in protecting its citizens from harm, it would adopt quite different policies from those now in force, policies that would involve regulation by the government, but not outright prohibition. The former policy has the virtues of aiding in consumer protection (a form of paternalism) as well as being less restrictive of individual liberty. In the end, however, Conrad's argument rests on the total ineffectiveness of present drug policy.

Precepts of Intrinsic Value

This category is one of the vaguest and least well-charted areas of philosophical thought. Put very generally, a thing is held to have intrinsic value if it is thought to be worthwhile in itself, good for its own sake, or inherently valuable. (The reverse holds for things held to have intrinsic disvalue—they are inherently evil, etc.) The concept of intrinsic value is usually contrasted with that of instrumental value—things held to be good because of some further desirable end or state of affairs to which they lead or to which they contribute. Another way to characterize this pair of value concepts is in terms of means and ends. Things having instrumental value are good as a means; those with intrinsic value are sought as ends in themselves.

The chapter in this volume that focuses most on questions of intrinsic value is the one by Dan Brock. Indeed, pleasure is one of the few candidates philosophers have nominated for intrinsic goods. Others are health, knowledge (for its own sake), and happiness, where that is viewed as something different from (or "higher than") pleasure. This is not the place to rehearse the various theories of intrinsic value and the challenges posed by some thinkers (John Dewey, for one) to the idea that there is even a meaningful distinction to be made between intrinsic and extrinsic value, or value as means and as ends. Instead, it would repay the reader to look again at Brock's careful analysis throughout his chapter; this section will be confined to a very brief look at places in the other essays where concepts of intrinsic value or disvalue are invoked.

One point Brock makes in the course of his analysis is worth repeating, even without its supporting arguments, since it serves to rebut a presumption mentioned (but not argued for) in other chapters. Brock writes: "It is highly unlikely that the extreme opponent of the use of drugs for pleasure can make out a plausible case that pleasure from the use of drugs either has *no* positive intrinsic value *at all,* or has intrinsic *dis*value" (p. 93). If after reading the argument in support of this conclusion we agree with Brock, we can summarily dismiss the positions mentioned in passing by Michels and Neville regarding the "intrinsic evil" of nonmedical drugs.

Michels makes the point in connection with his criticism of physicians who underprescribe narcotic drugs for pain:

> . . . one of the most distressing side effects of the social con-
> cern with the use of narcotic drugs for pleasure has been the exten-
> sion of the general moral disapproval of these drugs in such a
> manner as to contaminate medical judgment, with the result that
> physicians tend now to underprescribe narcotics for patients with
> severe pain, as though there were something inherently evil about
> drugs that produce pleasure regardless of the setting or purpose of
> their use (p. 182).

And in a very different connection, Neville develops the
thought experiment referred to earlier, as follows:

> Suppose a genuine Soma drug is developed, one with no
> harmful effects and that is safe in dosage. Suppose also that
> equally benign drugs are developed to enhance perception,
> awareness, memory, and alacrity of thought. Only pharmacolog-
> ical Calvinists would object to these on intrinsic grounds (p. 72)

Brock's analysis is sufficient to show that the notion of drugs as
"inherently evil" is probably an incoherent one. Yet the fact
that no plausible argument could be fashioned in support of
any such claim about the intrinsic evil of pleasure-enhancing
drugs fails to prevent some people from using those notions as
if they made clear sense.

Since so little use is made in this book of any concepts explicitly noted as ones of intrinsic value or disvalue, it is reasonable to conclude that they cannot bear very closely on the ethical issues before us. However, a rather different source of
intrinsic value might be sought in the notion of "duties to oneself," which Thomas Murray examines. Murray describes an
Aristotelian version of duties to oneself. The Aristotelian notion does not fit the idea of duties to oneself being of intrinsic
value, since Aristotle's ethics is essentially teleological, or directed toward a goal or end-in-view, which is held to be
desirable.

Another philosopher, Immanul Kant, elaborated a notion
of duties to oneself that might be thought a candidate for intrinsic value. According to Kant, all duties (to others, as well as to
oneself) must be done for their own sake, not for the consequences they may bring or for any other purpose, however noble. To act morally, rational beings must act from the *motive* of
duty, not merely in accordance with what duty prescribes.
Since duties must be performed for their own sake, in order to
count as truly moral acts, it may thus seem as though they are

candidates for intrinsic value. This characterization of duty according to Kant's philosophy certainly qualifies his ethical theory as a nonconsequentialist one; but does it aid in our understanding of the concept of intrinsic value? Probably not. For one thing, all duties must be done for their own sake, not only duties to oneself. Furthermore, the notion of intrinsic value is usually thought to apply to the property of goodness, while another ethical category—moral rightness—applies to actions that comprise duties. It would be stretching matters beyond their usual philosophical interpretation to try to construe duties to oneself as candidates for intrinsic value, so we had best abandon the quest for precepts of intrinsic value that might be thought relevant to the use of pleasure-enhancing or performance-enhancing drugs.

The Drug Controversy: Conflict of Principles or Dispute About Facts?

In contrast to the popular debate over the use of nonmedical drugs by individuals and the proper role of the state, the essays in this volume do not address the issue as a simple conflict between individual freedom and government paternalism or legal moralism. To be sure, that clash of principles is addressed by a number of writers. Murray, for example, in examining one answer to the question "May athletes use drugs . . . ?" observes that "[w]e have a strong legal and moral tradition of individual liberty that proclaims the right to pursue our life plans in our own way, to take risks if we so desire, and within very broad limits, to do with our own bodies what we wish" (p. 118). He notes, quite correctly, that "[t]hose who see performance-enhancing drug use as the exercise of individual liberty are unmoved by the prospect of some harm" (p. 118). Yet Murray claims that we are mistaken to look at this principally as a matter of individual liberty, and he succeeds in showing the respects in which that view is both factually in error and rests on a mistaken ethical analysis, as well. Murray's own analysis relies on facts about sport as a social institution and on a deeper probing of the "freedoms" athletes allegedly possess. As a result, he rejects the oversimplified view that the salient ethical issue must be seen as a clash between the value of individual liberty on the part of drug users (athletes, in this case) and the values embedded in paternalistic efforts to restrict drug use for the good of the user.

A potential conflict of values might be inferred from the essay by Dan Brock arising out of his portrait of the good for persons. The elements in Brock's brief sketch of the good for persons (p. 103) are ones with which few would disagree. Yet even among all who concur in the belief that these elements do comprise an objective set of criteria, some disagreement is likely to emerge on the question of whether people should be left entirely free to avoid seeking the good for persons or to abandon that quest in favor of drug-induced euphoria. One can agree that these elements constitute the good for persons, and yet disagree on the issue of value priorities: the freedom to choose the lifestyle one wishes may take precedence over the values inherent in a particular lifestyle. Brock does not address this question of value priorities directly. Yet he is quite clear in emphasizing that the underlying question is an empirical one, not something to be resolved by philosophical inquiry alone. The matter to be determined by empirical investigation is whether the use of recreational drugs necessarily interferes with individuals' pursuit of the good for persons in the way the "popular image" of the user of drugs for pleasure assumes (p. 103). If the popular image were accurate, Brock holds, it would indeed be true that the use of such drugs would interfere with important components of our good. But as Brock surmises, the popular image is "vastly oversimplified and substantially, probably largely, false" (p. 104). It comes down to a matter for factual inquiry, then, rather than being a question of philosophical and evaluative differences about the intrinsic value or disvalue of the use of drugs for pleasure.

It should be clear at this point in the discussion that the complexity of the issues can be traced at least as much to uncertainty and disagreements about the facts as to the multiplicity of value considerations surrounding the use of drugs as enhancers of pleasure or performance. A final, complicating factor in the public policy aspect of the debate is the existence of several different models for conceptualizing drug use and the formal and informal social controls operative in this domain. This chapter will conclude with a look at the "model muddle" in an attempt to see how those models contribute to the value concerns.

Drugs, Models, and Morals

The term "medical model" is both ambiguous and vague. It has multiple meanings, as Robert Michels points out, and still further meanings can be identified, meanings that shade into one another at the boundaries. Michels characterizes the concept as "a sociologic construct that describes the roles of physicians and patients, and the attitudes of others toward them" (p. 179). He adds a corollary: "the medical model has little to do with the organic etiology of various dysfunctions, and much more to do with illness and caretaker behaviors" (p. 179). The additional "medical models" Michels describes are the "public health model" and another variant that views the patient as "disabled rather than diseased, and focuses on support and rehabilitation rather than cure" (p. 182). Michels provides a useful schema, and I have no quarrel with the classification he proposes. But just as he notes the contributions of several features of these models to prevailing attitudes and social controls regarding the drug use of individuals, let me suggest that two additional variants of the medical model and their particular features further enrich our understanding of this cluster of issues. I refer to these as the "disease" model and the "physicians' control" model.

The disease model captures the by-now familiar phenomenon in our society of increasing "medicalization" of various forms of socially deviant behavior. Conceptualizing antisocial behavior as a disorder labeled sociopathy; calling drug addiction and alcoholism diseases; including tobacco smoking in a list of disorders certified in the APA Diagnostic and Statistical Manual (DSM-III), all might have struck physicians in an earlier era as odd. And, of course, there are well-known opponents of the tendency to medicalize socially deviant behavior even within the psychiatric profession today.

One reply, of course, is that current conceptions reflect scientific progress. But, it needs reminding that our understanding of these various behavioral disorders is not well informed by a widely accepted etiological account. Without trying to rebut or to justify applications of the disease model with its "medicalization" of diverse behaviors, moods, and mental states, I think that recognition of this variant of the medical model is important for understanding prevailing attitudes in medicine generally, and psychiatry, in particular, regarding drug use for pleasure. The fact that drugs act on the body and alter physiologic functioning in observable ways adds to the credibility of this picture and to the conceptions that underlie it. Think, for example, of the different responses to a person who says he smokes pot as a means of "escape" and another who reads thrillers or goes to horror movies for escape; or to one who acknowledges that he smokes dope to get high, for the pleasure it brings, in contrast to the person who says he listens to Mozart for five hours each night for the euphoric state it induces. The use of chemicals to alter one's mood or feelings renders the user subject to medical diagnosis and intervention, in part because physicians have expertise in matters pertaining to bodily functions (see Michels' discussion on p. 179).

These last considerations lead directly to my second variant of the medical model, the "physicians' control" model, which is closely linked to the sociologic construct Michels describes. This variant emphasizes the predominant medical control over various aspects of drug use in society. That control is justified by its proponents by reference to the expertise physicians have and layman lack in matters of psychopharmacology, dosage, toxicity, drug interactions, etc. The expertise of physicians is reinforced by their position of authority in matters medical, and also by political factors such as lobbying efforts in state and federal governments. Norman Zinberg observes that "current social policy designates all drug users as criminals, deviants, or even 'miscreants,' and in response to that policy the medical establishment designates all users as mentally disturbed" (p. 39).

Two brief remarks are worth making about the "physicians' control" variant of the medical model. The first is a reminder of the findings from Zinberg's own research, calling attention to the importance of set and setting, in addition to the pharmacologic effects of drugs, for understanding the differential reaction to psychoactive drugs that users experience. Set and setting may well be subjects for systematic scientific study, but they are surely not narrowly *medical* factors, the understanding of which is guaranteed by biomedical expertise. Zinberg has further demonstrated in his research the role that the experiences of drug users play in contributing to a sophisticated knowledge of acceptable dosages and the effects that combinations of drugs are likely to have. Many experienced drug users of addictive and nonaddictive substances are thus

able to titrate their doses more expertly than a straightfor-
wardly pharmacologic expertise could allow. Joined with the
operation of what Zinberg refers to as "informal social controls" governing all subcultures of drug users, it becomes clear
that expertise regarding drug use is not limited to biomedical
expertise. As important as biomedical expertise is for dealing
with the consequences of prolonged drug use and with over-
doses or adverse reactions, it does not warrant the predominant *formal* social control the medical profession as a whole exercises in this version of the medical model.

The second remark worth making in this connection is one
that emerges from a comparison between the American and
British experiences regarding the treatment of addicts, at least
until quite recently. The physicians' control model operated
with considerable power in Great Britain, allowing physicians
to maintain their patients addicted to opiates through doses administered in the physician's office and even by supplying patients with a take-home supply of heroin. This practice did not
represent a more lenient attitude toward heroin or towards addicts on the part of the British government. Rather, it reflected
the power and authority of the medical profession in England,
a power and authority that enabled persuasive members of the
profession to convince governmental authorities that since
drug addiction is a *disease,* the care and treatment of addicts is a
matter for physicians to handle, in the best medical tradition of
"clinical judgment." Thus until the last decade, when the de-
mographic picture of addiction changed markedly in England,
the "physician's control" version of the medical model operated even more powerfully and persuasively than it has in the
United States. The link between this model and the "disease
model" is also clearly shown through the British experience.

One variant of the medical model that Michels describes—
the "public health" model—is elaborated at greater length by
Bakalar and Grinspoon. The latter authors do not distinguish
between different candidates for the medical model, but the
particular conception they dwell on is the "infectious disease"
model. Cutting across the distinctions noted above, the fact
that drug use and abuse have been thought of according to a
model akin to infectious diseases helps to illuminate several aspects of the overall issue. For one thing, making drug use analogous to infectious diseases presupposes the appropriateness
of the expertise physicians have in matters related to diseases,

and this reinforces their power and authority in society's efforts to control the nonmedical use of drugs. In addition, as Bakalar and Grinspoon observe: "If drug abuse is a communicable disease, the drugs are a menace like the typhoid bacillus or the smallpox virus, the reasons for intervention become overwhelming" (p. 21). Since public health is one of the concerns of government, as well as a matter for the expertise of physicians, the infectious disease model of drug use and addiction affords an explanation of the mutually reinforcing roles of medicine and government.

Most attention has focused on the medical model for conceptualizing, explaining, rationalizing, or controlling the use of psychoactive drugs inside or outside of medical contexts. But the use of any one model is probably an oversimplification. Other so-called models employed in this domain are the "enforcement" model and the "moral" model for understanding drug addiction and social reactions to drug users. One can find the medical model and the enforcement models inextricably intertwined, as they have always been in the United States regarding heroin, and as they have increasingly become in Britain, now that heroin maintenance of addicts by physicians has been abandoned.

Similarly, historical studies reveal a mixture of disease and moral models at different times and places, one notable example being that of opium eating in the period between the eighteenth and early twentieth centuries. It is questionable whether any single model, taken alone, is sufficient for gaining a full understanding of the use of psychoactive drugs and attempts to control or regulate them. Bakalar and Grinspoon sum the matter up nicely:

> The vocabulary of public health medicine also permits a smooth transition from physical health to psychological and moral health and finally to social health: the "murderous epidemic" is crime and illness at once, without careful distinction. In this way consumer safety becomes mixed with morality and the two different kinds of justification reinforce each other. . . .
>
> Since this special public health problem includes psychological, social, and moral health, the use of drugs for pleasure is believed to present three kinds of threat to human welfare. It is an offense to morals or a danger to the social fabric; in some ways it is also like an epidemic disease; and in some ways it resembles the ignorant use of a dangerous instrument like a chainsaw. Social

attitudes and legal regulations conform to each of these three analogies in different ways, and each model in turn reinforces the others at their weak points to supply reasons for stricter controls (pp. 22–24).

This picture of the "mixed models" at work contributes to our understanding both of the intractability of efforts to solve the drug problem in our society, and also to the question of why social attitudes toward recreational drug users are at such variance with social tolerance toward other forms of risk-taking behavior in recreational activities.

We may despair at the sheer complexity of the ethical, social, political, and medical issues surrounding the use of drugs and attempts to arrive at sound public policy. But that complexity alone should not constitute a barrier to future efforts to address our society's "drug problem" with a more rational, effective approach than has been mounted to date. Legislators, administrators in the executive branch of government, physicians, and others involved in making drug policy and enforcing drug laws would do well to heed the recommendations made in several chapters in this volume. Reflecting on the implications of Bakalar and Grinspoon's analysis of social attitudes, accepting Conrad's conclusions about the abysmal failure of our drug policies, and implementing Zinberg's recommendations may not make the drug problem go away. But it would be a decidedly rational step in a more socially desirable and morally acceptable direction.

Reference

[1]John Rawls, *A Theory of Justice* (Cambridge, MA: Harvard University Press, 1971), p. 302

Subject Index

A

Abortion, and the right to privacy, 134–135
Alcohol, regulation of, 24–25
  compared to drug regulation, 25
  See also, Prohibition
Amphetamines, 110–112, 115
Aristotle, 119–121, 206
  eudaimonia, 119–121
Autonomy, 66–79, 119, 131, 193
  See also, Coerciveness, inherent; Privacy, constitutional right of; Psychotropic drugs, right to use; and Right to refuse treatment; Rights

B

Becker, Howard S., 8, 33–34
Body, control of, 140–142
  Roe v. Wade, 140

C

Chambers, Carl D., 56
Cocaine, 109, 112, 117
Coerciveness, inherent, 116, 119, 122–125, 197–200
Connelly, Harold, 109, 115
Connelly, Patricia, 115
Conservatism, 78–79
Constitutional right of privacy, *see* Privacy, constitutional right of

Consumer safety legislation, 15–17, 20–21, 204
  compared to pollution control laws, 16
  compared to seatbelt laws, 16–17

D

Decriminalization, drug use and, 41
DeMont, Rick, 118
Dewey, John, 205
The Drug Abuse Council, 29, 32, 62
Drug addiction, 5, 73–74
  medically based, 5
  socially based, 5
Drug control, 18, 21–22, 49–63, 144–151, 165–169, 175–185, 211–213
  See also, Consumer safety legislation
  history, 18, 49–56
  interests of State in, 144–151, 165–169
    nonpaternalistic interests, 145–146
    paternalistic interests, 146–148
  justified as public health intervention, 21–22, 211–213
  See also, Medical model, public health model

215

by medical community,
    175–185
problems of controlling
    prescription drugs,
    176–177
objections to, 148–151
principles vs facts, 207–208
regulation vs prohibition,
    37–39, 41, 56, 60–63
state role in, *see* Intervention
Drug use, *see also* Privacy,
constitutional right of, and drug
    use
acceptable levels, 37–39
as component of the good
    life, *see* Good life, place of
    drugs in
consequentialist arguments,
    see Moral principles,
    consequentialist
definition, 65–66
as epidemic disease, 21–22
legitimate vs illegitimate, 3–4
medical model, 2–3, 5–6
medical regulation, 23
natural vs unnatural, 120
pain-killing, 57–58
pleasure, *see* Pleasure
pleasure-enhancing, 59–61
risk-taking, 13–26, 187–188
    analogous to mountain
        climbing, 14–15
    analogous to use of
        chainsaw, 15–16
    positive vs negative
        consequences, 14–16
rituals, *see* Social controls
self-destructive behavior, 72
setting, 29, 34–35
social consequences of,
    195–196
social controls, 29–31, 35–37,
    41–44
social learning, 31–37
    See also Education, impact
        on drug use

in sports, *see* Performance-
    enabling drugs;
    Performance-enhancing
    drugs; Football
Drug user, popular image of,
    103–105, 197
Drug users, as scapegoats,
    77–78
Drugs, definition of, 178,
    184–185
Drugs used for pleasure,
    opponents of, 83–84, 90–97
    extreme, 92–93
    moderate, 93–97

E

Education, impact on drug use,
    16, 32
    See also, Drug use, social
        learning
Eighteenth Amendment, 51–53
    See also, Prohibition
Endorphins, 3–6, 88–89
Ethics, drug use and, 61–63
    See also, Good life; Moral
        principles; Pleasure

F

Fabric of society, *see* Drug use,
    social consequences of
Feuerbach, Al, 116
First Amendment, 137–139,
    148–149, 151, 159, 161–162,
    163–164
Football, 57–58, 110–111, 115
Fourteenth Amendment, 134,
    140–141, 143, 150, 159
    See also, Privacy,
        constitutional right of
"Frankenstein Factor," 1–2
Frenn, George, 110
Freud, Sigmund, 38

G

Good life, 98–105
  See also, Aristotle, *eudaimonia*
  common components of, 103
  place of drugs in, 98–100,
    102–105
  theories of, 100–103
    desire theory, 101
    perfectionist or ideal
      theory, 101–102
    pleasure or happiness
      theory, 100
Gutierrez-Noriega, Dr. Carlos,
  117

H

Harding, Wayne M., 29
The Hastings Center, 50, 57
Hedonism, 85–88, 92–97
Heroin addiction, as
    consequence of social
    setting, 34–35
Hoover, Herbert, 53
Hunt and Chambers, estimate
    of heroin users, 55–56

I

Inherent coerciveness, *see*
    Coerciveness, inherent
Intervention, 65–79
  government's role in, 76–77
  principle of, 66, 76
  by state, in drug use, 65–79,
    193–195, 203–204, 207–208

J

Jacobson, Richard C., 29

K

*Kaimowitz v. Dept. of Mental
    Health,* 68, 74
Kant, Immanuel, 206–207

Kaplan, John, 43
Klerman, Gerald L., 3

L

Laetrile, 135–136, 164, 167–169
  and right of privacy, 135–136
  right to use, 163, 167–169
Law enforcement, illicit drugs,
    54–56
Liberalism, 78–79, 194
Libertarianism, 192, 194, 200,
    203–204
Liberty, 118–123
  concept of, 118
  limits of, 118–123

M

Mandell, Arnold J., 110–111
Marijuana, 40–42, 149, 157,
    164–165, 167–169
Marxism, 79
McGlothlin and Arnold study,
    33
Medical model, 179–185,
    209–213
  characteristics of doctor in,
    179–181
  characteristics of patient in,
    180–181
  definition of, 179
  disease model, 209–212
  drug use and, 183–185
  enhancement model, 182–183
  normalizing and optimizing
    model, 182
  physicians' control model,
    210–211
  public health model, 181,
    211–213
  See also, Drug control,
    justified as public health
    intervention
Mental illness, correlated with
    drug use, 33–34, 39–40

Mill, John Stuart, 19, 21, 87,
   90–91, 94, 190
Mind–body problem, 89
Mind control, 137–142
   See also, Psychotropic drugs,
      right to use; and Right to
      refuse treatment
      cases—Cases:
      *Kaimowitz v. Dept. of Mental
         Health,* 139
      Olmstead v. United States,
         137–138
      *Paris Adult Theatre I v.
         Slaten,* 138
      Rogers v. Okin, 138–139, 161
      State v. Renfro, 164–165
Moore, Mark H., 43
Moral controversy, 188
Moral principles, 189–208
   consequentialist, 189–200
   intrinsic value, 205–208
   nonconsequentialist, 193,
      200–207
Morals, legislation of, 18–20
Motorcycle helmets, 149–150

N

Narcotics, contraband seizures
   of, 55–56
National Commission on Law
   Observance and
   Enforcement, *see*
   Wickersham Commission
National Commission on
   Marijuana and Drug Abuse,
   see Shafer Commission
National Institute on Drug
   Abuse, 38
Neurobiology, 1, 3
New York Drug Law of 1973, 62

P

Paternalism, 69–70, 72
   See also, Drug control,
      interests of State in;

Intervention, by state, in
   drug use
Patterns of drug consumption,
   27–29
Performance-enabling drugs,
   57–58
Performance-enhancing drugs,
   6–9, 57–61, 107–125, 185,
   197–200
   See also, Steroids,
      Amphetamines
   caffeine, 107–108
   cocaine, 109, 112, 117
   confused with therapeutic
      drugs, 117–118
   control of, 124–125
   ethics of using, 117–125
   justice, 199–200
   optimization vs
      normalization, 4, 7–8
Personhood, 137–142
   See also, Mind control; Body,
      control of
Pleasure, 83–105, 196–197,
   205–207
   and endorphins, 88–89
   extrinsic vs intrinsic value, 92
   higher and lower, 88, 90–97
   nature of, 85–90
   preference theory, 86–101
   property of conscious
      experience theory, 85–97
   quality and quantity, 93–97
Policy recommendations, illicit
   drugs, 43–44
President's Commission on
   Mental Health, 40, 192
Privacy, constitutional right of,
   129–151, 158–161, 164–165,
   201–202
   applied to children, 142,
      147–148, 151
   cases,
      Doe v. Bolton, 134
      Eisenstadt v. Baird, 133–134
      Griswold v. Connecticut,
         132–134, 141, 160

Meyer v. Nebraska, 131
Pierce. v. Society of Sisters, 131
Poe v. Ullman, 133
Privitera v. California, 136, 151
Ravin v. State, 141–142, 165
Roe v. Wade, 134–135, 160
Rutherford v. United States, 135–136, 151
Stanley v. Georgia, 134, 141
Union Pacific Railroad v. Botsford, 130
Whalen v. Roe 135
and drug use, 135–136, 144–151
history, 130–136
Supreme Court discussion of, 129–151, 160–161
Prohibition, 24–25, 49–55
Psychotropic drugs, 157–169
right to refuse, *see* Right to refuse treatment
right to use, 162–169

R

Rawls, John, 200
Remus, George, 52
Responsibility, 66–79
absence of coercion as condition of, 73–76
competency as condition of, 75
informed judgment, as condition of, 67–73
Right of privacy, *see* Privacy, constitutional right of
Right to refuse treatment, 157–169, 202
See also, Psychotropic drugs, right to use
constitutional bases, 160–166, 169
informed consent basis, 159–160, 162
Rennie v. Klein, 158

Rights, 200–203
constitutional, *see* First Amendment; Privacy, constitutional right of
liberty, 200–201
regulation of, 143–144
Risk, *see* Drug use, risk-taking
Robbins, Lee N., 28, 34–35
Rockefeller, John D., Jr., 52–53

S

Set, 29, 210–211
Setting, 29, 34–35, 210–211
Shafer commission, 39–41
Smart, J. J. C., 93
Social attitudes toward drug use, determinants of, 177–178
Social policy, 39–43
Sports, drug use and, *see* Performance-enhancing drugs
Steroids, anabolic, 109–125
See also, Performance-enhancing drugs
control of, 124–125
doses taken by athletes, 112–113
health effects, 112–115
impact on performance, 113–114
in Olympics, 109–110, 112, 117–118, 121–122
risks, 114–116, 121, 124–125
Szasz, Thomas, 43, 192

T

Tillitt, Malvern H., 51
Twenty-first Amendment, 51

U

United States Supreme Court, *see* Privacy, constitutional right of, Supreme Court discussion of

V

Vietnam, heroin addiction in,
  28, 34–36
Volstead Act, see Prohibition

W

Wickersham Commission,
  49–50
Williams, Roger, 147